GHOSTS OF PETRA

THE SPICE OF LIFE

Second Edition

Clifford D. Cope

Cope with Gravity

copes creations

6913 Duncan Circle
Fort Smith, AR. 72903
www.copescreations.net
(479)-629-5810

First published by Clifford D. Cope 7/30/2021

ISBN: 978-1-7376426-1-9 (e)
ISBN: 978-1-7376426-0-2 (sc)

Printed in the United States of America

Other books by author:

Desert MASH
Caleb
This Land is My Land
Trial of Billy Ranhalt
Otherworld
Ghost from Machu Picchu
Ghost of Saud

For my two kids, Kirby, and Bea, who had to
Put up with me while I wrote this novel
As well as to listen to me talk about ghosts.

Contents

sE= soul Energy

$(life=E+mc^2+sE)$ or $(E+mc^2+sE=Life)$

$(E=mc^2)$ or $(mc^2 =E)$

$\sum em=G$

Preface

If someday all religions are finally done away with, I believe we will then have more peace on earth. That is, we will have fewer Crusades, fewer Holocausts, no more witch hanging in Salem, no more 9/11s, no more Six-Day Wars, and no more Desert Storms. Without religion, I would say, we would have more peace on earth. There would also be many more people with money to spend on the family rather than the preacher's house and Mercedes.

However, would there still be a God? Would there still be an afterlife? Even without religion, would the purpose of life be easily defined? I say yes to all three questions. More importantly, the answer is easy, and costs nothing more than giving up some cultural inventions. It is a fact that all religions are man-made and embedded within us, most of the time from childhood—sometimes by force, sometimes from a strong fear and influence. However, as the world becomes smaller and more open-minded, we have the opportunity to explore our own consciousness, seeking out many unknown answers on our own. Now, am I saying that God is whatever one wants Him, She, or It to be? Nope. There is a lot more to God than that. Trying to understand God is like trying to understand the cosmos. After all, God put our one and only infinite universe into motion with the laws and forces of nature that cause and allow change within itself. From the tiniest elements to the celestial bodies' evolution, God causes things to happen. Things are evolving every day around us. Why be afraid to accept it? Wow, if you don't think evolution is happening, look at churches and religions. They have evolved from giving a goat for sacrifice to giving your credit-card number.

Where is the soul? This is a remarkably interesting question. This is like asking, "Is the universe expanding or not?" This is where it gets really important with God. Most would simply say it is in the mind. If that were the case, I would assume that when one is declared brain-dead by two or more neurosurgeons, the soul is gone. In such cases, diagnosed patients have snapped out of being declared brain-dead. What, did the soul return? These cases will be fought over for many years, as religion will be in court and in the streets. Or you can open up your consciousness and simply become more of an individual. That is what happened to the ghosts of Petra; they just became themselves without being ensnared by influence from atheist to Zen.

Chapter 1
Petra

"Good afternoon, and I welcome you to the little town here we call *Wadi Mousa*, as well as the *Cleopetra* Hotel. My name is Aali, how can I help you today?"

"Hello, I see you speak English very well. We came through Israel to Amman and then took a long bus ride here, and you are the first person we have seen who speaks English. I wish we could speak Arabic, but we can't … but anyway, we need a room for three nights," Sam informed the hotel manager and owner.

"I would guess you and your lovely wife are here to visit the old city of *Petra.* Here at the Cleopetra Hotel, your stay includes three meals each day. If you are going to be gone for lunch, as most are, please let me know the night before you go out, and I will have a sack lunch prepared for you," Aali said, and then with a whisper added, "A warning, you don't want to eat out on the trails with the *Bedouins*. You will see them here and there trying to sell you jewelry, clothes, food, and tours. My advice don't take anything from these people. They are goat herders sending their kids into the park to make extra money off the many tourists coming here."

"Okay, we will take your advice, but right now we would like a room—and we are hungry, so what time is dinner?" Sam said as Jill filled out the registration form and gave the manager a credit card.

"Dinner will be in about one hour. Put your things up and freshen up a bit, while your evening meal is being prepared. May I ask, are you here on a vacation? By the look of your backpacks, I would think you are explorers of the world. Where is it you are from?"

Jill spoke as Sam began looking around the combined dining room and lounge of the twenty-room hotel in the small town a couple hundred miles from the capital of Jordan. "We are graduate students from the New England area of the United States. We are just touring the Middle East on our summer break, and I am looking for a really neat site to work on my doctorate degree in anthropology/archeology."

"Ah, you have come to the greatest site in the world: Petra."

"We have heard that at every place we have been to for the past month. And yes, we have seen some of the neatest places in the world—the Dead Sea, Masada, and then there was old Jerusalem. What great places!"

"Wait until you see the buildings—that from the outside there is only one side, yet from the inside there are four sides..."

Jill turned to Aali, "What, what did you just say about the buildings? What does that mean?"

"Ah, you will see. Also, when you are there in the once-great amphitheater, sit and enjoy your lunch and think of what was happening in your view two thousand years ago," Aali said. He then hollered over his shoulder for dinner to be prepared.

Sam grabbed his backpack with one hand and Jill with the other. She grabbed her pack and the room key off the counter as she continued looking at Aali.

"I am ready for a shower. A private warm shower, that is—unless you want to share one with me," Sam said as he got to the room door, grinning at Jill as she tried unlocking the door. "I am kind of glad to be out of the cheap hostels where the shower stalls are worse than those, we had in my high school locker room."

"Uh, I don't know, I think I will wait and see what it looks like," Jill answered as she pushed the old wooden door open and then moved out of Sam's way so that he could go in first. He pulled his large pack in behind him, dropped it at the foot of the bed, and fell back on the bed. "Wow, this feels great," he said, closing his eyes.

"I see you found the nice shower," Jill said as she pulled her pack to the small closet area, opened it up, and started pulling things out. She dug a few things out and walked around the room, checking it over.

Sam responded with only a grin.

The shower was cozy but clean and inviting. While Sam napped, Jill took a long hot shower and got ready for dinner. As she gathered things up to be washed, she woke Sam from his nap. He roused himself slowly, got up, took his shower, and was ready for dinner in about ten minutes. They were joined at the large dining table by several other people who were staying at the hotel for the purpose of visiting the ancient ruins of Petra. Dinner was saffron-spiced rice and rotisserie-roasted lamb, along with a fresh salad of sliced tomatoes, cucumbers, and lettuce doused with a flavored olive oil and vinegar. Warm bowls of covered pita bread and various cheeses were placed along the center of the table to be shared by all. Several bottles of red and white wine were placed within reach. Jill and Sam absorbed tons of information on places they must see and some that were not so interesting. The places most dreaded were those one had to walk to for several hours up rocky twisting hills or pay for a strong-scented camel or donkey to ride. Jill was interested in the archeology work that was going on in the old *Byzantine* church and remains of the *Castle of the King's Daughter.* Sam was interested in the altar of Abraham on the highest peaks there. Talk continued into the late evening about the history of the mysterious hidden city. Jill and Sam finally called it quits and informed the manager to fix them a sack lunch and give them a wake-up call at four-thirty. They packed their small packs with bottled water, a camera, and a few snacks for their trip into the hilly desert, and then they sacked out for the night.

Four-thirty came early, but they were both ready within minutes. They ate a quick breakfast, grabbed their sack lunches, and headed for the main gate, which was only about five blocks away. There were very few people at the park entrance when it opened at five. They paid their fee and headed out to find the much talked about *Sig.* This was the high, natural, walled pathway into Petra. At the end of the half-mile-long narrow gorge, they were confronted with the famous *Al-Khazneh Farun* (Treasure of the Pharaoh). Seen in many movies, its true purpose is still unknown today, but most believe it was the tomb of *King Harith IV*. Jill was in awe when the carved facade came into view. She stared in wonder at the great-carved tomb for several minutes while Sam continued to walk toward the entrance. Inside, he found the walls and floor sported nothing, but body-size rectangular holes carved into the solid stone insides. They were all empty, leaving Sam quickly bored. As he exited the structure, he met Jill slowly walking in. "Well, seen that one, let's hit the hilly trails."

"Wait, Sam, let me have a chance to look around," Jill told him. "Sit down and read your book and stay out of stuff while I take some pictures and enjoy the site." When she had finished looking around, she found Sam a little ways off sitting by a sign at the trailhead to the mountaintop altars. She caught up to him, and they began their hike with the one or two other early birds headed into the main city area. As they struggled up the very narrow sandstone trail for about an hour, the sun began peering over the horizon and shone on the sheer cliff that they were hiking to. The sunshine would not radiate into the old city for a while, because it was set down on the valley floor surrounded by high steep walls. When Jill and Sam reached the platform that lay out in front of another facade structure, Sam took a seat on a large rock out in front while Jill went toward the carved building. Jill walked up to the wall, leaned back on it, and began unscrewing the cap to her water bottle.

"Whoa! Do you smell what I smell?" she said, sniffing the air.

"No, I don't, but do you see what I seeee?" Sam sang back as he looked up and down the only outside wall of the single-story house. It displayed alternating shades of tan, white, red, and brown horizontal stripes made from the mountain's stone. As Jill turned to look at the wall she was leaning on, a hidden voice sang out.

"Can you hear what I hearrr?" sang out a voice from inside.

Sam and Jill quickly glanced at each other with their eyes wide open. Jill put her hand to her mouth in a sign of silence. With a big smile on his face, Sam pointed to the one and only open doorway to the stone house they were looking at. They were early, but some other tourist must have beaten them in the climb. As they stared at the doorway, an old man appeared holding a wooden walking stick. He had a big smile on his face as he greeted the two young tourists.

"Welcome to Petra, I am Youssef. I like that song you began to sing. Do you know the rest of it? I see you have come a long ways off the normal path at an incredibly early time."

"Yes, we read in the park guidebook that this was a beautiful site to see as the sun came up over the mountain. Are you a tourist?" Sam asked.

"No, of course not, do I look like a tourist? I am a tourist guide. Are you looking for a guide?"

"No, but thank you for the offer, we have guidebooks for the places we visit," Jill said.

"Oh, but there is always more to see and to learn about a place or event than what one can find in the books."

"So, do you work here for the park or one of the tour companies out by the park entrance?" Sam asked.

"No, neither I live near here and have my herd of sheep, goats, and a couple camels enjoying their early breakfast just over the hills there," the man said as he pointed to the slight rise in the rocky hill.

He then added, "So what brings you two, way up here other than the sunshine? Would you like to see some of the great Nabataean structures that are not in the books? If so, I can show you them for just a small fee."

Sam thought for a few seconds and then said, "Yes we are out early, and yes we are off the normal path. We wanted to look at the places out away from the city. We also wanted to view the city from up here on the mountain. I hope we are not intruding on you. If so, we will go back down to the city."

Jill slowly began moving close to Sam while looking at the old man. She asked, "No, we don't know the words to the song. How did you know it was a real song? We did not intend for that to be a song; we were just being stupid, I guess you could say. What are you doing up here if your sheep are over the mountain?"

Youssef squatted down and reached behind the wall just inside the doorway. He brought his hand back out with a small clear glass of hot tea. "Ah, I was just up here sleeping and now having my early morning tea and a bit of breakfast. I sometimes bring my herd near here to stay the night. I sleep the night away here in one of these old tombs while my herd sleeps over the hill there. I have a kid watching them. During the day, I also try to make a few dollars extra giving tours to people like you while my herd grazes. Here, have a glass of tea, I have not drank from it. I have another here in my bag for you both," he told them as he got up from the doorway and walked out a bit holding the glass out for Sam as he nodded his head for Sam to come and get the glass. Sam stepped forward, reaching for the glass of tea, and Jill followed, taking the second glass. Youssef then returned to the doorway, reached inside, and pulled a small satchel bag out from behind the wall. He dug another glass out and sat down on the threshold where he could reach his teapot just out of sight. He poured himself a glassful, gave Sam and Jill a salute, and took a sip. Sam and Jill looked at each other and did the same.

Youssef then said as he took another slurp, "My village is too far to sometimes walk each day, so I at times just spend the night out here with the stars. As for the song, I know that it is an old Christian song about Christmas. I am one in the land of Islam faith, so I do not hear that one very often. I know the song because my father sang it to me; it is about a shepherd boy. However, it is just a song. Though, this building is not just a building. It as well as many others here are buildings that have been constructed that have only one side from the outside, yet four sides from the inside. Have you ever seen such a building before … or heard of any such structure?"

Sam looked at his tea, studying it a bit before putting it to his lips again, and then he looked at Jill. Jill looked at the old man as he poured himself another glass from a small brass teapot. He then got up and walked toward Jill, offering her a refill, but stopped short after seeing her full glass. As he turned around to head back to the doorway, Jill abruptly said, "I smelled the scent of a strong spice—like nutmeg or cinnamon or something like that. Right here! When we first got here, but I don't smell it now." Jill lifted her chin and walked back over to the wall she had been leaning on. She looked around the corner and continued to quietly sniff the air as she moved and looked around. "No, I cannot smell it now, but it was right around here that I smelled it. It was like walking into a bakery in old downtown Boston."

"That is because they are gone," Youssef said as he sat down and began eating a hard roll and drinking his tea.

Jill quickly turned to him with her hand out and asked, "They are gone? Who are they?"

Youssef smiled at her and said, "The smell of the spices, they are gone, maybe the breeze blew them away, or they just got up and went away."

"Ooookay, so where did the fragrance of the spice come from?" Jill asked as she moved to Sam, who had gone back to his stone bench.

Youssef walked behind her as she went to the bench, and then he turned and stood looking at the building, pointed at it, and said, "From there, that is where they came from."

Jill quickly glanced at the one and only wall from the outside then back at Youssef with a blank look. Sam just quietly sat and looked around as the old man talked. Youssef stood there dressed in a long blue-and-white robe with a turban on his head and a long walking staff in his hand. He walked over to the stone bench Sam was sitting on and sat down.

As he sipped his tea, he pointed at the building and said, "It comes from those places—from inside, and the walls and from—well, from all over. You see, the building is made from the mountain. Inside of the building there have been many tomb niches carved into the walls and floors. At one time, there were many laid to rest inside these buildings. But also, there was a time when some were not at rest. This is the building called the *Turkmaniyeh Tomb.* It was made in honor of the Nabataean god *Dushara* and designated his property you see over the doorway it tells about this god in the Nabataean language."

Jill walked over to the doorway and looked at the writing over the door. She then glanced inside at the darkness that filled the large room. Sam blurted out, "I do not smell anything over here but the tea, which I must say is great. I also can smell the sheep and goats, so they must be close by," he added as he sipped his tea.

The old man twisted sideways toward Sam, looked at him with a smile, and said, "You have a good nose,—for my sheep, they are half a kilometer from here over that hill."

Sam lowered his head and then quickly got up, went inside, and hollered out, "Yep, it has these old grave holes carved out of the stone walls and floors just like the other ones."

Jill followed him and began to explore the great room. It was not the first building they had explored. Just as they entered the park, they saw the famous treasure and several others in the lower valley as they climbed the steep trail. The buildings were not all carved out of the mountainside. There were also the broken remains of many stone-built buildings throughout the old city and scattered throughout the valley and mountain top. They would visit these later in the day after the made their visit to the highest structures. One other interesting structure intrigued Jill, which was the most complete one that stood in the middle of the old city. That was the castle. After looking around inside for a few minutes, they went back outside. When they stepped out of the door, they did not see the old man. Youssef was gone. His small backpack with his teapot and things was also gone. The two little tea glasses Jill and Sam had left on the stone bench were still where they had left them. Jill darted outside, began looking around, and called Youssef's name. Then she gently turned her head, listening for a response, but got none.

Sam came outside, walked around looking, and said, "I guess I should not have said what I did."

Jill looked at him with a questioning expression on her face and asked, "Said what? What did you say to him?"

"I simply said that I smelled sheep while he was sitting by me on the bench, but I did not really mean anything bad by that."

"Yes, you did smell the scent of sheep, but you didn't smell the sheep. Gosh, Sam, you just do not say things like that around people. You probably made him feel bad, and he left to go back to his herd."

"I guess we could go look for him and return his glasses." Sam went over to the bench, picked up the little glasses, and held them out, one in each hand, toward Jill. "Maybe we should just keep them; we could use them as shot glasses back home."

Jill walked over to Sam, took the two glasses from him, and said, "No, let's just leave them inside the doorway. Maybe he will find them tomorrow or another day. Let's finish walking the trail as the guidebook says to the other sites, and then we will go back down to the city and look at the buildings the archeologists are working on. That will take us the rest of the day. Tonight, at the Cleopetra Hotel, we can talk about what we are going to do tomorrow and whether or not we are going to stay another night."

"Okay, let's go," Sam said as he gave Jill the glasses and headed for the trail. Jill took the glasses and placed them inside the doorway in the spot where Youssef had his little teapot set over a small fire. She then looked out of the doorway toward Sam as he walked toward the trail. She dug down in her backpack and pulled out a card, rolled it slightly, and stuck it in one of the glasses. She then caught up with Sam. Sam was reading aloud from the tour guidebook as he walked down the path. From his reading, Jill gathered that the next point of interest was what was called the *Lion Triclinium* or *Lions Tomb*. This was a hypogeum—a structure that was partially underground, but also had the great one side carved into the mountain with two lions carved on each side of the one and only door. It could have been used to house the animals and gladiators that fought here in the city's amphitheater. It had the head of a Medusa carved above the door. They stopped and explored the inside and outside. Sam soon settled on the outside steps and continued reading as Jill studied the carvings. This trip was Jill's idea; she had just completed her undergraduate degree in anthropology at the University of Pennsylvania and was starting her doctorate degree the upcoming fall when they returned. Sam was a bit different. He had just completed his undergraduate degree in math and was working on his doctorate in biophysics and nanotech at MIT. They both were intrigued by the latest of the latest in the world of science. They also had become intrigued by each other. Jill had to stop and study every little drawing or carving they came across and highlight everything they had seen in her own guidebook. She wanted to work on a site in the Mideast but was not sure exactly where. Sam wanted to work in *nanotechnology*. His dream was to build the world's smallest computer or one that would act like a human, as were several leading modern technologist around the world were.

As Jill knelt down in order to look at one of the lions, she suddenly smelled the aroma of a spice. This time it was different from before. She stopped Sam from reading and told him to come over by her and smell.

"Do you smell cinnamon? It was right here by the door," Jill asked Sam as he came toward her.

He moved his chin up, walked around Jill, and said, "Well—yes, yes, I smelled cinnamon as I walked up the steps to you, but I do not smell it any longer. I do not smell it here." Sam turned and went back down the three steps and walked around a bit with his chin held high. Jill just stood at the doorway and watched Sam.

"I am not kidding, Sam. I really did smell cinnamon like I smelled something over at that other building, but now it is gone. This is strange. Just where is it coming from? I bet you don't believe me, do you?" she asked Sam with an innocent smile.

As Sam slung his backpack over his shoulder and picked up his guidebook, he said, "Yes, my dear, I believe you. I also smelled something cinnamon-y. It smelled like that little cinnamon-roll shop in the mall back home. But maybe it's just us or what we had for breakfast, or maybe there are cinnamon trees or other spice trees around here, and the breeze is blowing it to us." Sam then started out on the trail again.

Jill jumped from the steps and joined Sam as she looked behind at the tomb. She grabbed Sam's hand and said, "Cinnamon trees do not grow here, Sam. Nothing, hardly, grows here."

Sam looked around, agreed with Jill, and started reading from his book as he quickly walked to the next and last structure on the trail. That was the *Al-Deir* structure. This structure was carved up on the plateau where they could look down on the central part of the old city of Petra referred to as the Petra site or park. The nearby small modern city also called Petra, Jordan. Officially it goes by the name, Wadi Mousa. Then there is the old city referred to as Little Petra which is kind of a Bedouin town outside the Petra Park. Sam and Jill were staying at the Cleopetra Hotel which was within walking distance of the Petra Park entrance gate. This large carved structure was given the name Al-Deir by modern Arabs because it looked like a monastery or holy place. There was no evidence as to what its exact purpose was, as was the case with the whole city. The carved building was about 150 feet tall and almost as wide. Out in front of it was a large open area said to have been used for large gatherings of people. There were no four-sided structures along the steep trail to this place or around it. Al-Deir was also a building that had only one side from the outside but four sides from the inside. There were three large rectangular openings carved out of the mountain, appearing as windows—but only one allows for a view into the mountain. That was for only a few feet, and then it ended. One could not ever look out, except from the one and only door. Inside there were two stairways, each one leading to a small niche. It was said that this structure may have been built for King Obodas I, who ruled Petra in the first century before Christ. After looking around the open plaza areas as well as many of the structures, Sam and Jill had one more climb. That was to the top of the mountain where there was a large water reservoir carved into the rock. It still holds water today but has several cracks in its walls that allow the water to slowly seep out, feeding a lush garden below it and off to the side of Al-Deir. After seeing one of many large carved water cisterns, Sam and Jill took the narrow steep path down to the garden to enjoy the lunch the hotel owner made and packed for them. They spread a blanket out against the shady wall to get out of the hot sun. They began to see other tourists far off in the distance at the beginning of the trail begin their ascending journey, just out the back of the Petra museum by the amphitheater. At a steady walk, it took a couple of hours for someone to get to the high place of worship or social gatherings. A small stream flowed out of the tributary valley they were in that led down into the main valley. The garden was referred to as the oasis of the mountains by the hotel owner/manager, who was the one who told Jill and Sam that it was a perfect place to have lunch. After a quite long lunch, they climbed back up the little tributary wall to view the sun setting on the Al-Deir facade. They went to the cave opening across the plaza and found a spot in the shade along with about thirty other people who had climbed the trail to see the sunshine directly on the mountain sculpture. After about ten minutes of waiting for the sun to set, Jill was alerted to the scent of spice. She elbowed Sam.

He raised his chin, turned his head from side to side, gave a bit of an alarmed look at Jill, and then said, "That's it, that's the scent I smelled back at the Lions Tomb."

Jill gave him a great big smile and said, "See, I told you so, but why does it come and go?"

She looked across Sam and gave a small wave at a group of people looking at them. They looked back at her and waved back. Jill asked them, "Did you smell the scent of spices like cinnamon or nutmeg, or like you would smell in a bakery?" The man sitting next to them shook his head and said something in Arabic to Sam. Then a man sitting behind them spoke out for the man shaking his head, translating what Jill had said and what she and Sam were talking about. The man in the group smiled at Sam and Jill and said something. The translator stepped in again and said the man said he did not smell anything but would like to smell a bakery—but that would take another two-hour walk back to town. The translator added that he also smelled something like a spice, but it must be because they are hungry. The scent went away, and they did not encounter it again as they walked back down the long steep valley trail that brought them out near the Sig entrance into Petra. From there they went back to the main street of the park. Jill had to see one more site before they returned to the hotel, which was the Petra Church. It was on the main street not far from the castle. It was another subject of the several Old Testament relations that had been found or talked about in Petra. The church was that of the Byzantine Empire, who were the very first Christians, most likely. It was discovered as a church in 1990 and excavated from 1992 to about 1998. In 1993, there were about 150 papyrus scrolls discovered in the church. They were writings about the history of a great family who once lived there. The scrolls are being studied still today. What Jill found most fascinating was the great number of mosaics that covered the walls and aisles. They were all finely sculpted pictures of animals, flowers, people, and structures. They seemed to tell a story. Each of the mosaics was made from tiny squares of marble. The different colors of marble showed many fruits and animals from faraway locations. There were animals like elephants, large lions, and tigers. There were fruits and vegetables from hundreds of miles away. The church was evidence that the Nabataean's were great travelers or provided a stopping place for traders from far away.

Jill zeroed in on one mosaic in particular. That was one depicting the great *Qsar Al-Bint Temple,* also known as the Castle of the King's Daughter. Even today, it remains one of the greatest centers of attention on the valley floor. In the mosaic, the plaza out in front of the castle, is covered with tiles with six sides all interlaid to cover hidden aqueducts as well as secret tunnels for performers, gladiators, and animals. The mosaic showed a giant pineapple lay out on this plaza. However, today, this can hardly be seen because the paint has been worn away by 2000 years of weather and footsteps, yet the stones remain in place. Jill almost wore the camera out on these mosaics as well as the remains of the castle.

As the late evening approached, they headed straight to the park entrance and then to their hotel on the edge of the modern town of Petra. The owner/manager, Aali, was sitting on the porch drinking a beer when they arrived. He jumped up, greeted them, and asked how their trip was for the day and if they enjoyed their lunch. He then motioned for them to sit down and take a break while he went inside to get them a glass of wine or anything else. They dropped their small backpacks on the floor and collapsed in the large lounge chairs. Aali soon returned with a couple of tall, fluted glasses of white wine. Sam and Jill sat back and began enjoying their wine with the owner, who retook his seat. He said to them, "A light afternoon meal is being prepared for you and will be ready in a few minutes."

Jill sat up and placed her glass on the table, looked at Aali, and asked if he had ever smelled the scent of spices anywhere in the old city the many times he had been there.

He turned to look out into the street and replied, "No, I have not, but I have heard stories from one or two people who said they smelled something strange in various places, but not just one place or always in the same place. I think it is because people read about Petra as being a shelter or resting place for the early spice-trade caravans that traveled through here two thousand years ago, but I do not think any of this spice would have survived until today."

"Were the people you heard the stories from tourists or local people who live here?" Jill asked.

"Oh, I have heard the stories from both, but they are just stories. In this country of Jordan, there are many old stories and legends, because we are the center of human civilization. We are just south of the first city in the world, the city of Ur …"

"Okay, so you have heard from tourists in different years or seasons that they have smelled the scent of spice but never seen any," Jill asked again.

"Oh yes, I believe so. Please allow me to see about your meal."

Aali stood up and excused himself to go get their afternoon snack. As he went through the door, Sam hollered for him to bring the bottle of wine when he returned.

Jill said, "Why is it that we are not the only ones who have smelled this glimpse of spice, yet it cannot be found? I mean, as we saw in old Jerusalem near the hostel, we stayed in. There were those little spice booths along with all the other stuff in markets with large baskets of all kinds of spices. You could not miss them."

Sam looked at her as he sat back and enjoyed his last bit of wine and said, "But Jill, my dear. Those were big baskets of spice in a small, crowded hallway. Up there around those buildings, we did not see great baskets of spice other than that in the foodstuffs they were selling along the trails. I have also seen some selling little bottles of stuff. Maybe that is spice. Maybe it is one of those little bottles that have been broken or spilled. What food they do not sell, they just throw behind the rocks. That is why you see so many stray cats around these hills, I bet."

Jill replied, "Nobody is going to sell spices up there. Yes, we have seen kids and ladies all around with their little blankets of modern-made '*ancient jewelry*' and clothes, but not jars or baskets of spices."

Aali returned with a bottle of wine and the kitchen servant carrying a large platter of various meats, yellow and white cheeses, olives, and sliced vegetables and dates. The servant also placed a basket of various hard rolls and breads on the table. He then unfolded a cloth napkin and laid it in their laps. The servant was excused, and Jill asked Aali if he would continue to tell his story of Petra.

He took his seat and continued telling them that the ancient city of Petra was once a thriving city built within the low mountains of northern Jordan several thousand years ago. It was a central waypoint in the long trade route between Egypt and Damascus across the great desert of the Persian Gulf area. The city of Petra could be approached from various directions, but it was built in a valley of great rock formations that allowed fortress-style architecture within the mountains. There was no easy route entering the city through the natural steep high cliffs of sandstone surrounding it. These routes allowed entrance to the city through selective corridors cut into the hills through time by nature's flow of water, giving the founding Nabataeans added protection. The city was once home to about thirty thousand Nabataeans in its prime of life until it was taken over by the Roman Empire in 106 AD. It soon began to decline in growth and popularity. The Roman population dropped very sharply shortly after it occupied the city. The Bedouins today say that many of the Romans began to dislike their leaders and fled to the surrounding deserts to live among the Bedouins.

"However, this has been likely proven wrong by the archeology work here at the park," Aali added. "They have said that the city declined due to the great earthquakes that accrued here just after the Romans took over."

Jill interrupted by asking, "So how did they prove what happened?"

"Oh, I do not know, but these people have been working here for years and have learned and recorded many findings. They are scientists, so how can they be wrong?" Aali said.

He then went on, explaining that the word Petra means "crevice in the rock," going back to the popular Greek language. It was most likely given that name because of the cracks and fissures throughout the region—especially the great fissure leading into the city from the east, called the *Sig*, meaning the shaft. Throughout the mountains surrounding the city were many water cisterns and dams made by the Nabataeans in order to collect water during the limited rain that fell on the land. The people also had constructed many open water ducts as well as closed conduit waterways, bringing the city water from miles away. Once, a castle occupied the central part of the city, with a large amphitheater and hundreds of other dwellings surrounding it. There were many stores for food and supplies to get the people through each dry season. The spices that the traveling traders carried were stored in the buildings carved into the mountains alongside the tomb structures. That is why some of the one-sided building had the tomb carved niches in them while others were just open rooms. These tight carved structures kept the people and trade goods safe from bad weather and robbers. Here in this land, the spices were like gold and silver in other lands. The most different thing about Petra from any other town in the world or in the same region was that most of the buildings were built into the mountain. Only the buildings' facades could be seen from the outside, which were all hand-carved within the dimensions of the *golden rectangle* like the rectangle framing the Mona Lisa painting. The city remained under Roman rule only for a few years. It then fell under the Byzantine Empire, which tried to establish a rule also until finally the city became somewhat abandoned and returned to the local people. In the heart of the city near the amphitheater stood the *Castle of the King's Daughter*, or *Qsar Al-Bint Temple*. It was almost totally built from yellow limestone blocks. The ground around it was the only ground paved with large six-sided stones. Today, one only sees parts of the castle standing. Legend says it was built for the beautiful princess *Katrina* as well as a fortress for her family and other important people of the city. Her family was the last rulers of the Nabataeans. The owner had just started to tell them about the legend of the princess when they were interrupted.

From across, the street came a loud "Hello, my friends. I see you have made it back to your hotel."

"Oh my God, it is him—the tea man," Jill said as she set her wine glass down and waved at Youssef.

Quickly, Aali said, "Oh, watch out for this old Bedouin. He will want to give you a tour of the old city. My advice is to let me find you a better guide. He is just a crazy old goat herdsman." He then got up and went inside his family hotel. Jill stood up and waved for Youssef to come over and join them. The old man looked up and down the street and then slowly walked over to the hotel. He climbed the three steps and gave a slight bow to Sam and Jill.

"Please take a seat and have a glass of wine and some cheese and crackers. We owe you a drink at least," Jill offered as she pulled a chair out for him.

Youssef sat down, pulled his little bag into his lap, dug around in it a moment, and retrieved a small tea glass. He held it out for Jill to pour him some wine. He then settled back in his chair, taking a sip, and saying, "Ah, it is such a lovely evening, did you enjoy your walk to see some of the other structures away from the main city? There are so many others, and they are all so beautiful."

"What do you mean, there are many others, are there others that are more interesting?" Jill asked while Sam cut open another hard roll and placed a couple of slices of meat on it, followed by a couple of slices of white cheese and a little mustard. As he took a big bite, he offered Youssef something to eat by pointing to the tray. Jill sat on the edge of her chair looking at Youssef, waiting for an answer.

"Yes, of course, there are other sites of interest, and of course depending on your interest. Today there are a few archeologists here working on this town hoping to discover something or learn what the purpose of this here old town was. They say that the structures are falling apart or flaking off from the outside because the water is washing salts and minerals out of the stone. Then, when the water evaporates away, it leaves crystals just beneath the stones' surface that swell each time new rain comes. But that is not always the case," Youssef explained as he made himself a sandwich.

Jill turned closer and, directly facing Youssef, asked, "Why did we smell the scent of spices? Why did it go away? Have you … ever smelled the scent of spices around the old city?"

Youssef took a bite of his sandwich and a long slow sip of wine. He rolled his eyes toward Sam, who had stopped eating his sandwich and was looking at Youssef, awaiting the answer. Youssef turned toward Jill and softly said, "Yes, I have … I smell the fragrances each day. They come and go, but they never go away. They just go to another place. They …"

"Why do you keep saying they? The scent of a spice does not just pick itself up and float away. Who are *they*?" Jill interrupted.

Youssef continued as he took another slow sip. "These do … You see, they are just not one fragrance because there are many different spices to be found in this country. These spices you smelled here today, well, they are like somebody. They are like a ghost of spice. You see, they are like a lost soul that was left behind when the person died. They are a soul locked in a spice. You see, there once was a people living here whose souls were not allowed to go to the Otherworld, as in their afterlife. Their soul was trapped in the essence of various spices. You have smelled what they slept in and what they lived in while they waited to die, trapped in some of those structures you saw today. Kind of like the smell of various animals that some people carry because they work around those kinds of animals yet are not dead," Youssef answered as he looked at Sam.

Sam looked at Youssef, raised his hands in surrender, and said, "I am sorry for what I said this morning. I did not mean anything insulting to you. I was just questioning Jill about her smelling cinnamon, and I really did smell sheep or goats. But I did not mean anything bad about it."

Youssef raised his hand, closed his eyes, and slightly bowed his head to Sam. "Your apology is accepted, and yes, you did smell goats and sheep. Each day I have to carry a young lamb or two to keep them up with the herd so that they are not left behind as some people leave behind loved ones, family, and friends. Now that you have made contact with some of them, maybe you would like to meet them?"

"What do you mean, we made contact with them, and if we would like to meet them?" Jill asked.

"I mean—maybe you would like to go and talk with those that you sensed and see them as you are seeing and talking with me right now."

Jill broke in, saying, "Wait, wait, wait, this is getting a bit out of hand. What do you mean? Like as in go out, meet a basket of spice, and have a chat or something? I am sorry, Mr. Youssef, but that is just a bit out of whack. I don't think we would be interested."

Youssef shrugged his shoulders, bowed his head, and said, "You made contact with them when you smelled the spice. You were standing right in their presence. If you would like to go see them, please … you are welcome to meet me at the *Attouf Ridge* in the early morning just after the sun comes up. You will find it in your guidebook as *Abraham's Altar*." Youssef cleaned his small glass and restored it to his bag. He also pulled two glasses out of his bag, placed them on the table, and slid them over toward Jill, saying, "These I left for you; I see you also left them for me. Now I return them to you. I hope to see you in the morning. Now I must go, for my sons await me just outside the town." Youssef got up, bowed to Sam and Jill, and then went down the steps. As he waved goodbye, he added, "Oh, and bring your things, because I think you will want to spend the night." Sam stood and thanked Youssef for stopping by and said it was nice talking to him. Jill just sat and looked at Youssef as he left. Sam returned to his chair and his eating.

Jill sat for a few moments in silence watching Youssef disappear into the darkness and then said, as she continued to stare into the empty street, "The owner is right—he is a crazy sheepherder. I think I drank too much of this wine or whatever it is." She studied the glass she held in one hand and twirled her blond curly hair in the other, and then she finished drinking the last bit of wine.

Sam sat back in his chair and said to her, "Jill, my dear, it's not the wine you're drinking. This all started up on the mountain this morning. When we did smell the spice and run into Youssef. Maybe he was waiting for someone and not just resting there. How he found us again, I do not know, but he did. Now he is giving us the chance to go chase a ghost story. Do we dare check out of this place tomorrow and go to Abraham's Altar? That place was in our plan, so I say we go tomorrow—but do we take our things?"

The hotel owner came out and asked how their little meeting was and if they would like anything more to eat or drink.

Jill looked at Sam and said, "No, thank you, we have had enough to eat and drink. We are going to retire to bed early tonight, so we will not be eating the late dinner. We will be checking out early in the morning, because we are going to tour the valley portion tomorrow and catch a bus at the back gate."

Sam got up with a blank look on his face, finished his wine, made one more sandwich to go, and headed for the door. As Jill got up to leave, she was stopped by the owner who held his palm out, warning them again. "I am sorry for interfering in your business, but I must say again that I do not think you should get involved with this Bedouin. He will tell you crazy stories of Petra and the Nabataean people. He tells the stories because he knows that is what tourists want to hear. There is no truth to these stories, and he is going to ask you for money to be your guide."

Jill said to him, "Don't worry, we need to just see a couple more things here then get back to Amman. We want to see the excavation that is currently going on, and then we can go to the back gate and be on our way to Amman this evening. We will not pay the old man anything, as we have not paid any guides so far in our weeks of travel."

Jill started to go inside but Sam stopped her, saying, "Hey, Jill, let's walk down the street to the store. I need to get some film before we leave tomorrow." As he went down the steps, he said to the owner, "We are good at guiding ourselves wherever we go; if we have not paid in full, please let us know tonight because we will leave very early." Jill followed Sam down the steps and down the sidewalk.

When they were out of voice range, Jill said to Sam, "Uh, Sam, we have a digital camera. We do not need film."

"I know, but I wanted to get away from there. I think he just wants money also and would like for us to stay there forever, as long as we are paying. We need to get a few snacks like bread, cheese, lunchmeat, and some fruit to take with us. We also should get a few bottles of water at least for a day or so because I am not sure what we are going to learn tomorrow. As for the old goat herder, if he sounds too bullshitty tomorrow at the altar, we will just go on. Nevertheless, I want to see what he is talking about. How about you, do you feel up to this?"

Jill grabbed his hand, pulled him toward her smiling face, and put a big kiss right on his lips. She said, "Yes, I am ready to see what he is going to show us. After all, I am an anthropologist and archeologist by degree now, and I need all the experience I can get with the study of people."

Sam laughed a bit as he hugged her around the waist and said, "Yeah, right, my dear, I want to see you explain your visit with the ghosts of Petra, Jordan, in your first graduate class this fall. Wow, you might be the first *ghost anthropologist* in the world. It's a good thing you did not get your undergraduate in archeology alone, because I think it would be hard to dig up ghosts."

Jill playfully pushed Sam into the street and said, "Well, one does not have to dig up a ghost to find one, now does one? And you, what are you going to do, figure out a way to use nuclear radiation on them? But you know, I always wanted to go check out a ghost story, so now is our chance."

When they got to the store, they bought a few items to fill their backpacks as they did every few days. They had been gone from the states for several weeks now, staying mostly in cheap hostels and out of expensive restaurants. In Jerusalem, Tel Aviv, and a couple of other sites in Israel, hostels were about ten dollars a night and full of young travelers like themselves. These places required your own bedding, so they were prepared to camp out or sleep in a cave if necessary. While in the store, Sam asked the cashier if he had ever heard of any ghost stories about Petra, or if he had ever known of any guides that take tourists to see ghost places.

The man quietly added up the bill. He looked at Sam and Jill and said, "That will be thirty-one dinars, or twenty-two dollars and fifty cents." Sam handed him the money, gathered up their things, and turned for the door. The cashier then said, "This is something you should not talk about here. This is a myth among the Nabataeans, which are a few old Bedouins who call themselves Nabataeans. The government as well as many others here do not care for this talk." He then looked back down at his money drawer. Sam said thank you and started out the door. The cashier then added, "If you should smell the fragrance of a spice out there somewhere, I think it would be wise for you to run. Run to where there are many people. At night, when you sleep, if you dream that you are in the midst of evergreen, which is the aroma of *frankincense* or *myrrh*, wake yourself, get up, and run. It is said that the spirit of the *Djin* roam about in the hills around Petra. They will try to appease you with their aroma. When they have you in their grasp, they then will suffocate you with their strong fragrance. Be careful when you talk around the smell of spice." The man then turned and went to the back room.

Sam and Jill looked at each other and walked quietly back to the hotel, where they packed their things for an early morning departure. They both took a long hot bath and went to bed. But before Jill could fall asleep, she got back up, walked over to the window, and looked out. She reached up and felt the lock. She then went back to bed and fell asleep as Sam snored quietly.

Chapter 2
Ghosts

Early the next morning, they finished packing and went to check out. Aali was not there. The hotel servant checked them out, and they headed off toward Old Petra again. Because they went through the main gate with their large backpacks, they were required to sign a paper stating that they were not coming back through this gate. There was another gate at the north end of the park that they said they would be leaving through. Camping was not allowed in the park, so when people entered with large backpacks, the park guards wanted them out at closing time. When they reached the Castle of the Pharaoh's Daughter, called Qsar Al-Bint Temple by the officials, they stopped to look at the archeological dig that was going on there. An archeologist from the United States had been conducting an ongoing excavation for the past ten years. The team was just beginning to go to work and was having a morning job assignment and briefing when Jill and Sam approached them. Jill went straight to the lead archeologist, introduced herself, and explained what she was doing there. They were invited to sit and listen to the morning briefing and, when that was done, the archeologist would talk to them for a few minutes. After announcing the day's job assignments, the lead archeologist gave everyone a briefing on the latest examinations that had been going on to help stop the weather erosion of the many structures there. She explained how this destruction was the result of salt contamination that was in the sandstone structures. Each year, when the area received heavy rainfall and flooded the area, the salt was dissolved and then pushed to the outer layers of the stone. When the water evaporated, the salt re-crystallized, swelled, and caused the outer edge of the stone to flake off. This short class went on for about thirty minutes, elaborating on how many of the buildings and tomb walls were slowly being destroyed. She elaborated on how she and her colleagues were formulating prevention. At the end of the class, the summer students and paying volunteers went to work in their assigned areas under a graduate student's supervision. As they headed off, Jill raised her hand in question.

The professor said, "Yes, one question from our visiting student, and then we must go to work."

Jill asked, "If you would please, two questions: first, are those six-sided stones that make the outside plaza floor here also those that are shown in the mosaics on the floor of the church? I mean, the mosaic in the church looks like a model of the castle with a giant pineapple painted on the plaza floor. You cannot see but a few small splotches of dye or paint, but the shape of the stones look like the skin of a pineapple."

The archeologist professor answered, "These two structures have nothing in common because they are of two different time periods a couple hundred years apart. Okay, one more before we need to go to work."

"Thank you! Now on the information board over there I have read where research and several preventions have been going on with this problem of the structures' surfaces disintegrating for many years now. Do you think the cause might be from another source? Also, why is it only happening to certain structures?"

"Well, first of all, we should not debate the credibility of our ability here and the research we have been conducting. We are very well certain of the cause and the effect. We also have proven that the structures being destroyed are ones that are taking on water. We will soon have a better solution to stop the destruction that is taking place than that which they are applying at this time. We just need funding to further our study and prevention procurements. Okay now, let's all go to work."

The professor picked up her clipboard and walked to her office tent as the students went to their work location. Jill turned to Sam and said, "That, that …"

"Don't even say it. Let's go, we have a long climb to the altar," Sam said as he grabbed his large backpack and slung it on his shoulders. He then grabbed Jill's, holding the straps open for her to back into. Sam had to nudge the backpack up to Jill as she continued to glare at the tent.

"Let's go, Blondie, you're just not her type."

Jill rustled into her backpack, turned to Sam, and said, "Don't call me that again. I told you that before." She then took off at a fast walk for the trail, leaving Sam a bit behind with a grin on his face.

"Sorry, but I just had to get your attention. That always works."

After their visit, they located the rocky steep trail and headed for Abraham's Altar. They questioned each other about why Youssef wanted to meet them on the highest point there was and not at the building where they all first met. Abraham's Altar was a small gathering place that was known to have been used for animal sacrifices, and where Abraham supposedly was to have sacrificed his son, Isaac. The altar was a square of about twenty feet cut into the top of the mountain about eight inches. The local teachers say that it was made to honor the Nabataean deities before Christianity had been created. The male god, Dushara, was sometimes referred to as the god of the sun and the lord of life. The altar of Abraham had a small holding area where a black stone was placed and where it is said blood was poured. There are several channels carved into the altar floor where the blood supposedly flowed to Dushara.

There were two carved obelisks to the side of the altar, one representing Dushara, the other Al' Uzza, the feminine god of the Nabataeans. The moon and Venus symbolized her. She was the goddess of water and fertility. After about an hour's climb, they reached the summit. They looked around and quietly called out for Youssef but got no answer. The two obelisks, like the large rectangular altar of Abraham, were carved out of the mountain by workers literally cutting the mountaintop down about twenty feet, leaving two spears of stone and several other structures. Sam and Jill walked around the edge looking over the steep sides. The only thing visible was a Bedouin village off to the south several miles away.

Jill looked toward the village, placed her hands on her cheeks, and yelled out, "Youssef."

After a couple of yells, she turned to Sam, who was sitting on one of the few seats that had been carved out of the mountain, and asked, "Do you think we are in the right place?"

"Yes, we are in the right place where he told us to be. However, you know he did say he would see us just after the sun got up. Maybe we're late, and he went back home or went down the other trail looking for us. Let's just sit and wait for him." They sat around for a while and then walked around looking at the old ruins of a small fort that the Nabataeans had built to protect the altar. They had already made plans to spend the night but could not stay at the altar because there were tourists who came there to see the site. At the closing of the park, a park official made the rounds on every trail, pushing tourists on to exit the park. So, they decided to walk down a small southward ravine trail where they could see an oasis of green trees. It looked like it was on a trail leading to the village they could see in the distant. The oasis they could see was only a five-minute walk from the altar, so they could return there occasionally to see if Youssef were there. The oasis was similar to the one they had rested in the day before. It was loaded with many small bushes covered with pink flowers, a few varieties of cactus, and a few other small bushes. There was a small stream of water coming out of the rocks just below the mountain rim where the altar was. It flowed right through the middle of the small oasis. Of all the walking they had done in Jordan, this was only the second spot of flowing spring water they had seen. The trail seemed to continue onward toward the village that was out of site for them now. They found a nice spot near the spring running through the oasis to pitch their tent and set up for the night. They would be out of sight from the altar, so the guard could not see them. Before darkness set in, they heated up some water to cook some ramen noodles with chicken broth. They sliced up some cheese to go along with a bit of sliced camel meat and the challah bread.

It seemed that every time they were eating or drinking, they met up with Youssef. This was no exception. Just as they were enjoying their meal, Youssef came walking up the trail. Sam and Jill just sat quietly as he came walking toward them and said, "I see you made it. I had forgotten about the park officials and thought they might have caught you trying to camp out at the altar. I thought maybe you were sent out of the park at the closing time by the authorities, so I was returning to the front gate and to your hotel to see if you were still interested in meeting my friends."

"Yes, we are interested and have waited for you since morning," Jill said to Youssef.

"So where are the spicy ghosts?" Sam added.

"First, we must have tea. Please, continue to enjoy your meal. I was at the altar as the sun arose but did not see you after waiting for several hours. I thought you might have decided not to come, so I went back to my herd. But now I am here, and I see you are still interested."

"Did they come with you?" Jill asked.

"Do you smell their fragrance?" Youssef answered.

"No, we do not smell them," Sam said, looking all around.

"Will we be able to see them?" he asked.

"Yes," Youssef said as he pulled his teapot and small glass out of his bag.

"May I use your little camp stove there to fix some tea?" he asked, pointing to the small propane camp stove sitting idle.

"Yes, please use it and tell us about the ghosts of Petra while your water is heating up," Sam told him.

"First, we must have tea. This is Arab tradition, and then we will talk."

"Will we be able to talk with them and see them here tonight?" Jill asked.

Youssef sat cross-legged as he prepared tea and slowly bowed his head down and to the right as in a dance curtsey to Jill, and then he quietly said … "Yes."

Sam and Jill sat looking at each other, waiting for something to happen and, at the same time, both wondering if they were being led into a trap. So far, they had not been hit up for money, but they still wondered if that was going to happen.

Therefore, Jill had to ask another question before the tea was ready. "So, do you want money from us? Are you going to charge us to meet these ghosts?" she asked, sitting up on her knees.

Youssef raised his hand, held up one finger, and waved it back and forth, giving a no for the answer. He put out three glasses and slowly poured tea into each one. He handed one to Sam and one to Jill. He raised his glass as in a salute. Sam and Jill follow suit, and they all took a drink.

"You will meet three of them here tonight. You are then welcome to meet some of the others if you would like tomorrow," Youssef said, looking across the spring into dusk.

"Are they harmful?" Sam asked.

"Now is a bit too late to ask such a question. Do you not think?" Youssef asked.

"Yes, it is a bit too late, but will you answer the question?"

"Yes, they can be harmful, just as you can be. But if they wanted to harm you, they would have done that long ago."

"Where do they live? Do they live here in the mountains or in that village below in the valley?" Jill asked.

"Most of the time, they live in and around the ruins of what is called, Little Petra here on the south side of the mountain, away from the tourists and their old homes in the main part of the old city. There they have to be with bad memories and the great number of tourists."

Jill quickly jumped to her feet, walked a few yards out to where she could see the few lights in the village below, and said, "I want to see them." Then she loudly called into the early darkness, "Hello out there, ghosts of Petra! Where are youuu? You can come out now!" She then turned around and looked at Sam and Youssef, flung her arms out, spilling her tea, and said, "Well, I don't see or hear them. I don't even smell them, Mr. Youssef."

Sam slapped his thigh while laughing and said to her, "God, Jill, I cannot believe what you are doing. Hellfire, why don't you use your cell phone and call them. What do you think they are going to do, just pop right out like they do on TV?"

Suddenly she turned back around toward the spring flowing out of the mountainside and saw the handrail of the small footbridge move. At the same time, she stepped backward and said, "Whoa, what's the smell, wow, it is strong."

"It's cinnamon," Youssef said. "It is Saud; he is standing right before you, just across the stream. Ask him to be seen."

Jill stepped back toward Sam as he got up, holding his arms out as if to catch Jill.

Youssef remained sitting in front of his teapot and poured himself another glassful. "You see, each of them holds the fragrance of the spice they perished in. That is how I know that is Saud. Do you also smell clove and nutmeg? They are just to your left, but that is not their name."

"Yes, I do smell something different. But I do not see anything," Sam said as he starred at the bridge while trying to pry Jill's hands and arms from around his waist.

Jill softly said, "Please make yourself seen, we want to see you. Please …"

Suddenly yet faintly, three areas in front of them began to glitter. The shapes of three human figures began to appear. The one in the front had its hand on the rail.

When they became completely formed, Sam burst out, "Wow, you look like people wrapped in Saran wrap and then sprinkled with glitter like we used on Halloween." The three of them floated across the small log bridge yet appeared as if walking.

Youssef said, "Meet Saud in front of you, Haifa to his right, and Latif to her right."

They each raised their hand in a hello wave and said hello in English with a thick accent. Jill continued stepping backward, dragging Sam a bit. They both stared at the three ghosts in silence, as the fragrance went away.

"Say hello," Youssef said, looking at Sam and Jill behind him.

Jill raised her hand slowly and said hello to them. Sam did the same and asked, "Do you talk, and do you speak English?"

"Yes, of course we talk. Did you not hear us say hello in English? Yes, most of us speak English. We have listened to many different languages over the past two thousand years, so we have learned a few." Saud said.

Jill stepped forward a bit with her hand out, offering a handshake. Saud extended his hand out and took Jill's. Just as he softly clasped her hand, Jill pulled it back and gasped as she was shocked by a light tingle.

She grabbed her hand with the other, rubbed it, and asked, "What was that I felt?"

"That was me you felt," Saud said with a grin on his face. "I am sorry I frightened you, but I did not mean to. I should have told you we carry a bit of a tingle or shock, but it is harmless. It is just the way we are when we become visible." As Haifa and Latif crossed over, Saud nonchalantly pointed down and to his right. Latif moved his hand to Haifa, steering her to avoid a large wet spot on the ground. They walked around Saud's left side, avoiding the large damp spot, and came face to face with Sam.

Jill said, "Oh, I am all right, it was just a bit of a shock to not only see or hear a ghost, but, whoa, to touch one was quite a shock. Of all the ghost stories, I have seen on television and read about, I have never heard of touching a ghost. So, do you go through walls or people?"

"No, not really, we kind of get around just like you do. To touch a live human is as shocking to us as when you felt me. We are not always, as you see us now, and we are the ones controlling who sees us by becoming visible. When we are like we are now as you see us, anyone can see us," Saud said.

Sam moved around to view the sides of them and said, "Wow, I cannot see through you as if you're solid, but it appears that one could. How did you just appear?"

Saud looked at Sam and said, "We will get into that later. First, we must just get acquainted."

Jill walked back up close and said, "I see you as a real live person, but you also remind me of the chandelier that hangs in my house back home. I always was fascinated with the Strass Crystal prisms and ornaments that made up most of the chandelier. You glitter as if you were made up of all of the prisms as they hang together, with the shape of a—human, a man human."

"And me?" asked Haifa.

Jill looked over at Haifa, with the fingertips of her right hand still clenched in her left hand. "Oh, you look the same but with a different tint and …"

Sam stepped up beside Jill and added, "You are a girl, I mean a lady, that is easy to see, and a lot younger than Saud, but about the same age as Latif."

Jill elbowed Sam and then continued to stare at the three figures.

"Why do we remind you of a chandelier?" Latif asked.

"When I was seven, along with the great help of my dad we bought Mom this beautiful hanging chandelier for her birthday. It was huge. It had several hundred prisms of different shapes and balls cut from lead crystal hanging throughout the curved golden arms. We had it put in our dining room. The great chain that it hung from had to be twenty foot in order to get down from the vaulted ceiling. Early the next morning, my mother was fixing me waffles with whipped cream and strawberries for breakfast. As she was fixing mine, I soon noticed different shapes of rainbows appearing and then disappearing all over the room. The room had great windows filling two walls from the floor to where the wall met the pitch of the roof. I had never seen so many spots of rainbows before. I chased them around the room like butterflies but could not catch any. My mom told me they were happy spirits that came through the chandelier we bought her. Shortly after breakfast, the happy spirits always went away; that meant it was time to go to school. But they came back each day unless it was stormy. I never could sleep late like a lot of kids because I just had to go see those lines and spots of red, orange, yellow, green, blue, indigo, and violet. They danced around on the granite bar where I always ate breakfast with my mom …"

Sam put his arm around Jill as she looked down at the ground in silences and quickly asked, "How many of you are there? And do you live in families and have children? Do you just live like normal people?"

"We live with some of our family, and there are children that live among us. They are children left behind, as we were. But we do not make families here or additions to families, if that is what you are asking. We have lost members of our family just as you do today, yet we cannot be with them as someday you will be with yours," the young lady, Haifa, said as she looked at Jill.

"Please come sit with us and talk to these traveling students," Youssef said.

Sam put his arm over Jill's shoulder, moved her gently aside, and waved his arm in the direction of their camp kitchen. The three ghosts came closer and appeared to take a seat on the ground near Youssef as he drank his tea. Sam could not take his eyes off them, as when one pays close attention to a magician doing a trick to see if you could figure out how he did it. "Would you like some cheese and crackers or something to drink?" Sam asked.

"No, thank you for the kind offer, but we do not eat or drink. We are here mostly in spirit. We require energy just as you do yet get it differently." Saud said to Sam.

Sam quickly said, "I apologize for my way with words, but I am so curious about you and how you exist. What do you mean by 'mostly in spirit,' what else are you?"

"This is true about the mathematician Sam here, he does have a peculiar way with words, but I do not think he can lie," Youssef added.

Saud answered, "We are the soul that was left behind—that is, we are the soul of a person who died here many years ago that did not go to the Otherworld as souls do when their body expires. Now we are stuck here. When we become inactive or sleep, we become powerless souls attached to spice."

Sam and Jill had eaten most of their evening meal and put what was left to the side out of sight. They felt uneasy about being able to eat in front of others who could not. They also felt uneasy just being in the presence of something that would be referred to as paranormal. For Sam, that was something that did not pass the scientific method test. For Jill, that was something that fell into the pseudo-anthropology category. The ghosts, as Youssef called them, could be smelled when they were invisible. They also could be seen when they made themselves visible—but then could not be smelled. They could be talked to as if they were just another group of tourists. Now that the ice had been broken, there were many questions.

Jill asked, "So where do you live, and how many of you are there?"

Latif, the youngest looking of the three, said, "We live throughout these mountains and in the village below where Youssef and his family live. There are only a few of us; for some it has taken many years to escape through the pores of the stone, so I am not sure how many of us there are. Some have …"

Saud quickly interrupted, "Some of our people have no knowledge of who was trapped and who was not. The love of my life, for example, my wife and children, I have not seen since the day I was imprisoned in one of the spice rooms."

Latif then added, "And the love of my life I also have not seen since that day. We were forced apart by the new rulers. She was not put in the stone prisons like we were, she …"

Jill broke in. "Wait, wait, wait a sec. This is getting a bit confusing, and I see Youssef is being incredibly quiet. Why did you want us to meet the ghosts?"

All was silent for a few moments, and then Youssef said, "This all began when you and your friend came upon a couple of ghosts. To me, you appeared as if you were interested in meeting them the morning we met. I gave you that opportunity to just go on to see as all the other tourists come to see. You could have walked away. You continued to seek out information, and I gave you the opportunity to meet the ghosts of Petra. So here we are."

Jill replied, "Of course, Youssef, we were interested in the unusual fragrance that we stumbled on out here in the middle of nowhere. We had no idea that it was the scent of people we could not see—or ghosts, as I see now. Yes, we became interested, as we are in many things of this world; after all we are students. However, what I am asking now is, why you are so interested in introducing us?"

"Are you spending the night here with us, and are they also?" Sam asked as he pointed to Youssef and then over to the three ghosts.

"No, we will not be spending the night with you. We will return to *Quweira*, where you can find us tomorrow, if you would like," Youssef said as he sipped his tea, looking at Sam.

Jill asked again, "You still have not answered my question. Why were you so interested in introducing us?" she asked Youssef as she stared at him.

Youssef answered, "My friends, we will talk about that tomorrow. For now, let us just become familiar with each other and introduce ourselves. Tell these wonderful souls who you are and why you are here in their old home."

They all sat silent again for a few moments, and then Jill opened up, introducing herself as a graduate student at University of Pennsylvania in Philadelphia. She explained a bit about her interest in the people of past cultures and said she was going to complete her doctorate degree in medical anthropology/archeology. She had met Sam at a medical research convention that was held on her school's campus their junior year. Together, she and Sam shared a common interest in the blossoming field of human consciousness coping with modern technology. Sam was also just beginning his graduate studies at MIT in Cambridge, Massachusetts, doing his doctorate degree in biophysics. They had been good friends and shared a similar interest ever since they met.

Sam listened to Jill and agreed with everything she was saying by nodding his head up and down gently as he looked down at his empty tea glass and back at Haifa as she also looked at him intermittently. Jill joked around about how Sam's biggest adventure was to build or be involved with building an 'anthrocomputer' combining artificial intelligence along with artificial life. As Jill continued talking about their interests and studies, Youssef began packing his tea utensils in his bag. Saud and Latif began watching Sam as he drew pictures in the sand as he sat cross-legged. Haifa, getting restless, stood up and walked over to the beginning of the spring where it flowed from the rocks. As she stood there looking at the small stream, Sam stopped drawing in the sand and gazed at her.

She reminded him of his high-school prom. He was on the prom committee to organize and decorate the old army officers all club at the nearby closed army base in his hometown. The prom theme was "Castles in the Sky." Throughout the building, they covered the walls with cardboard painted as if they were castle walls. In every doorway, window, and sitting booth, the opening was shielded with crystal paper. It was transparent, but it appeared as though one was looking through a crystal blanket with a sky-blue tint. Of course, there were many other table arrangements and balloons, but he remembered the twinkly Star Gossamer paper the most. Looking out the windows and going through the doors gave one a view of misty shadows and a sense you might have in a dream. As he looked at Haifa, he thought he could see through her. After all, she was a ghost, but he could not see through her. Sam got up as Jill continued to talk and approached Haifa as she looked down at the small stream of water.

As Sam got near her, she turned to him, saying, "The water is endless. It comes from the sky, filters through the mountains, and gathers somewhere in there. Then it seeks an exit and flows down, until someday it finds a way out of the mountain. Some of it continues to the ocean, the lowest gathering spot, or is evaporated back up into the sky, where it is given another chance to go to the ocean if it did not make it before."

"Is that like you?" Sam asked.

"Yes, but I have not been given another chance."

Jill stopped talking about her and Sam's background and changed the subject.

"Will Sam and I be safe staying here tonight?" Jill asked Youssef as she glanced at Sam and then back to Youssef.

"Yes, sleep well and come to my town tomorrow. Just ask where I live, and you will find me. There you can learn more. Maybe you can help the ghosts, or you can just see Little Petra, which most tourists do not see. You will have a nice comfortable place to stay at my house, rather than up here on the hard ground," Youssef said as he got up.

As Jill also got up to say goodnight to the four, she asked, "After we visit with the ghosts again, and they have told us about themselves, can we then go from your town by taxi or bus to Amman? We did not get to see the museum there by the great amphitheater. We will need to go there in a day or so and then back home. Right, Sam?" she said loudly.

"Yes, we will be there before noon," Sam replied in a distracted voice.

Youssef smiled a bit and said, "Yes, you can catch a taxi in my town of Quweira. You will be comfortable there visiting with my family. For now, it looks as if it might rain, therefore we must go. The ghosts can smell rain like you can smell spices. We must go. Tomorrow I will take you up the valley on this other side of the mountains where you will see the neglected Little Petra."

The three ghosts and Youssef went down the trail as Sam and Jill watched them as far as the darkness allowed. Just before they were overtaken by the darkness, Haifa turned around and held her palm out in a still open hand wave. Sam waved back and stared outward.

Jill looked at him and said, "God, Sam, they are ghosts. We are encountering a paranormal event that no one will believe, and I don't quite know how to take it. Maybe it's time to head back to the hotel and get the heck out of here." Sam just stared out into dark space. Jill softly yelled, "Yo, Sam, wake-uuup, snap out of it. I can't believe it; you have become possessed by a ghost—well, really, possessed with a ghost.

"Sorry, I just cannot believe what I just saw. I don't believe it. We have just seen three ghosts and …"

"Sam, do you want to stay here the night, or do you think we should return back to the hotel?" Jill asked as she walked back to the tent and pulled the sleeping bag out.

"Yes, we will have to stay here; it's just too far and rugged to go back to the hotel in the dark. Why are you pulling the sleeping bag out of the tent?"

"God, Sam, there is no way I am sleeping in there thinking there might be a ghost lurking around out here. I want to be able to see them if they come back in the night. Heck, old Youssef said there are others, so, where are they?"

"I agree. Just remember you can only see them if they want you too."

They both looked around their camp with the glow of the little camp stove, gathered up their headlamps, and sat down at the door of their tent.

Softly and very quietly, Sam said, "I think we will be fine here. I think if they wanted to harm us, they would have already done so. I also wonder why Youssef wanted to introduce us to them. I mean, there are thousands of tourists going through here each year, and he picked us. Are there other picked people out there like us? He mentioned the word 'help,' just what did he mean by that? That is what possessed me—and I want to know what they are made of, just what is a ghost?"

"Or did we pick him?" Jill said to Sam as she walked back over to the stream. She looked at the stream and then at the wet spot between her and the bridge.

"Maybe it is because we are students or from another country or kind of like other young people who just take off and tour the world. They made it a big point to cross the bridge and not take the stone-hopping crosswalk just above the bridge as most others do at this trail crossing. Also, notice how they made it a point to not walk across the wet spot here," Jill said as she pointed down in front of her.

"Well, Jill, it's a water puddle. You and everyone else, except a halfwit, would walk around it," Sam told her as she studied the area.

She then said, "We are going to move the tent over here and put it on the wet spot. Maybe it will also make it cooler. That way we can sleep better tonight, because I have a feeling tomorrow is going to be a long day in Quweira."

Sam looked at Jill and slowly shook his head as he walked toward the tent and said, "I thought we were sleeping outside of the tent." He pulled the eight little tent stakes out of the ground, pulled the tent over the wet spot, and re-inserted the stakes. They moved the rest of their things over closer to the tent, climbed in their zipped-together sleeping bags, and talked. They tossed and turned for an hour or so and talked about their ghostly encounter of the day.

Finally, Sam grabbed Jill around the waist and whispered, "We could *huchecuche.*"

"No, Sam, not now, not tonight, I mean, yes, I would love to *huchecuche,* but maybe *they* are outside listening or watching. God, Sam, can you not think of anything more serious right now? Do you have your headlamp in your hand?"

"Yes, to all of the above. I am ready for the close encounter," Sam growled.

"Yeah, right, the only close encounter you had tonight was with the *girl* ghost. Are you still thinking about her?"

"Ah, well, I guess a little, I also think they do not walk as we think they do. Actually, I believe they float and just portray that they are walking. But they do not like water." Sam answered normally.

Jill quickly reached down, grabbed Sam's wrist lying over her belly button, and pulled it toward him. She then turned on her side away from Sam, saying, "God, Sam, even if you are thinking about her, you do not say you are, to a girl. Now go to sleep!"

"But that would be lying, I just cannot do that. But I am sorry; wake me if you hear anything."

"Okay then, just fib to me a bit sometimes. I will wake you. Let's not talk anymore and get some sleep."

They awoke early the next day just as daylight naturally illuminated their tent. As Jill stared at the tent ceiling, she asked, "Well, we are still here. Did I have a strange dream, or did we meet three ghosts last night?

Sam sat up quickly. "I think both, but then I guess I had the same dream. Were the ghosts you saw really sparkly and were they wearing clothes that just seemed to blend with their bodies?"

"Yes, that was them. They walked, talked, and sat as if they were like us. Well really they just kind of float walked and sat, I think," Jill said as she continued staring at the tent ceiling.

Sam got out of the sleeping bag and began getting dressed. He unzipped and flung the door open, and then he said, "I think Saud was about fifty or sixty years of age, Latif and Haifa were about twenty. Do you think they were brother and sister?"

"Yes, I would say they were about our age. I don't think they are related."

Just after he crawled outside, Sam turned around on his hands and knees, looked into the tent, and asked, "So what was it like to touch one? I have got to go see them again and touch one. Could you not hold onto him or was it that weird?"

Jill replied, "You would have to ask that question. I am sure you are going to have to touch one, and I bet I know which one that will be."

Sam growled at Jill like a wolf and entered the tent a bit, to be met with a hidden foot chasing him back out the door.

"Well, it really was not a bad shock," Jill said. "It was just a tingle. His hand was warm, not cold as I have heard in ghost stories and movies. It felt as if it was any normal hand but with a warm tingle. It felt like your first kiss, warm and tingling. Moreover, up close he looked like one of those glass balls you see in electronic or specialty stores. You know, those electric globes that have the blue electrical streaks zapping around inside of them. Just before I touched his hand, I could see more glitter flowing toward my hand as I got closer to his, as if his hand wanted mine."

Sam backed out of the tent whining and whimpering, and then he began digging around in his backpack, saying, "Yeah, they seem almost human like us, but with a glow—yet they do not eat or drink, and that's really sad. However, they must get their energy from something, or they would not survive, I mean they would not exist if they had no energy. Everything in this world we live in has some form of energy, and it must be gotten from somewhere. They have got to be made up of something, or you would not have felt them. You felt their energy … but there is more to them than that."

"Sam, they are a ghost, a spirit. You know, something that exists only in a vision or sound. They are the spirit of the dead. Let's not try to apply physics to them like you try to do with everything," Jill said.

"Maybe they are not dead. You touched one. Was that not more than a phantom or a vision? Maybe our soul or spirit has matter and energy—after all, everything else in the cosmos has that. Also, we have learned that everything in our *one and only* great universe has energy, and everything has matter and each is always in the process of converting to the other. Oh, but I will not bore you with all that again, but I want to go see them. I want to get a hold of one and examine him … or her. This is something we cannot pass up. We could be the first humans to ever really encounter real live ghosts," Sam said as he fired up his little stove and put a pan of water on for some breakfast tea. Then he dug around in his large backpack for some pita bread to go along with the fruit jams and peanut butter he had laid out on the large flat rock by the tent.

Jill got up and dressed, and then she started walking around outside, looking things over. "They don't make tracks," Jill said as she walked back over the small log bridge.

Sam looked at her as he spread a large white towel over another bench-like rock and began laying out breakfast. He said, "Well, where they were walking, I don't think anybody could leave tracks, the ground is so hard. Can you see if you are leaving tracks?"

Jill stopped and looked behind her and around the camp and said, "No, I cannot see my tracks. Except where the water overflowed the stream from the rain two days ago. I see your tracks and mine, but that's all, and I do see where Youssef sat. However, they did not walk in the wet spot; they acted like a cat walking around water. I guess we will see if they leave tracks today in the village. They are so interesting. Just think—they are a walking, living history book. Just think what they have been through in the past two thousand years. I cannot wait to see them again also. They are like a living book, a book of the truth, not a book of opinion."

"Well, get over here and eat some breakfast so we can head out for the town."

Jill went over to Sam, sat down, and fixed herself a pita sandwich of peanut butter and jam as Sam had done. They ate and talked about how strange their adventure had turned out and that they would soon be interviewing ghosts in real life. Sam finished his sandwich and tea, and then he began taking down the tent. He stuffed the tent and sleeping bags in his pack.

Jill was stuffing the food and the cooking and eating utensils in hers and laying out their bath kit and a couple of towels when a loud voice came from behind them toward Petra.

"I see we have found you."

Jill jumped and screamed in a bit of surprise. Sam also turned quickly and grabbed the bundle of tent poles as a club. As they looked up the trail, they saw two men looking down at them. They were two of the park authorities or guards. The one in the front came walking down the trail and asked, "You are Sam and Jill, yes? At the main gate you wrote that you were not exiting there but that you were exiting the north gate … But you did not."

Jill and Sam looked at each other, wondering what to say. Soon Sam spoke out. "Yes, sir, that is who we are, but we got a little lost looking around, and when it started getting dark, we felt it would be best to just camp rather than getting lost more in the dark."

The park guards walked right into Jill and Sam's little camp, looking over their stuff as they walked around slowly.

"I see that you have … and now that you are packing, where is it that you are heading?"

Sam looked over at Jill and then quietly said, "We were heading for the exit and then on to Amman."

"And which way is that?"

"It is that way," Sam, said as he pointed toward the trail that they all had just come from.

"You are going to Amman from here?" the guard asked as he sat down on the rock and watched Sam pack his bag in slow motion.

"Yes, we wanted to visit the great museum there and then fly back home."

Just before the guard stood up, he reached out and picked up two BlackBerry phones laying out on the rock. "What it this, BlackBerry? What, no iPhone? I see the battery is extremely low. I will hold these and have them recharged when we get to Petra." As he held the phones, he walked up the trail a bit, slapping his leg with a horse crop in his hand. He then sat down on another rock, looked at Sam and then Jill, and asked,

"Very well, we will wait for you and help you back down the mountain to Petra. There we can help you find a taxi or a bus to Amman at the park office. There are many taxis and buses there for you to take to Amman."

Jill finished packing her bag and said, "Oh, we thank you, but we won't need any help. We have a map here and will find the north gate today. That way, we will not bother you, and we can also look around at some of the things we have missed here."

The guard sitting on the rock was looking down at the crop in his hand as he held it up toward his face. He then turned his head up and toward Jill a little so that his eyes were pushing the top of his sockets in order to see her, and then he softly said, "No, we will go with you." He then asked, "And did you visit with anyone while you were here?"

Again, Sam and Jill exchanged glances at each other as Sam slowly said, "Oh, we have met so many neat people here, and they have all been so nice telling us what they have seen and good spots to go see and what restaurant to go eat at in town and …"

"Here, I said!"

"Well, just before it got really dark, an old sheepherder walked through here."

"And …" the main guard said as his assistant continued to walk around with his hands on his hips, his right hand resting on a sidearm pistol fastened to his belt.

Sam shook his head and said, "No, we have been alone all night." He finished packing his bag and joined Jill waiting on the rock bench.

Raul, the main guard, said something to his assistant in Arabic. The other guard replied with a couple of words and then headed off down the trail that they had just come from. Raul stood up and waved his hand in a low swing motion, signaling Sam and Jill to follow the other man. As they walked by Raul, shifting their packs around and getting them more in place, Raul said, "Ah, such lovely hair you have, my lady. In this country, someone like you should keep those curly blond locks covered as a Muslim lady would."

"But I am not Muslim. And it's okay for a lady in your country to not wear a veil, isn't it?"

"Ah yes, my dear, you are correct. However, when a man sees those long golden curls, it reminds him of gold. But I must say with no offense that polished gold jewelry blends so much more beautifully with strands of glistening dark hair."

Sam turned around, reaching back to take Jill's hand, and asked, "Do we just follow the guard?"

"Yes, of course, he will lead you straight to the office."

After a few minutes of silence, and lumbering down the trail, Raul asked, "So how was the old man, the sheepherder?"

"He was nice," Sam said.

"So, what did he have to say?"

"He told us about the people who once lived here and the names of some of the buildings." Silence fell upon them again for a few minutes.

Raul then said, "So this old man, he spoke to you in English, did he?"

Jill jerked Sam's hand and then said, "Yes, he spoke very broken English, and he only talked to us a few minutes. Then he walked on."

"And his name, did he tell you his name?"

"Uh … It was Saud; yes, I think it was Saud or something like that," Jill answered as she got up beside Sam as the trail widened.

"That is a strange name. I have never heard that name here before, and I know of very few people here that speak English. Ah, and for the Bedouins, I know of only one. Are you sure his name was not … maybe Youssef?"

Jill squeezed Sam's hand, Sam squeezed hers back, and she said, "I am not sure, the Arab names are so hard to say, and I get them wrong. It could have been that name; I am not sure."

The guard replied with a little laugh and said, "We will see, we will soon see."

They walked the remainder of the way in silence. When they reached the castle, the archeology crew was hard at work, scattered throughout the ruins. They walked by the office tent and could see the lead site archeologist standing behind her tent desk. She looked up to see Jill and Sam being escorted to the office. She came out from behind her desk and to the doorway and lightly hollered, "You need to stay on the trails when you are in an unfamiliar area. It is good that they found you and that you are safe. Good luck in your studies." She then turned and went back inside her office.

Jill barked out, "We were always on the trail, for your info. I hope the disintegrating walls fall on your tent and your students have to dig you out."

"That is right, you were always on the trail, and you knew exactly where you were going, just not exactly what you were getting into. Also, I must say, you are a feisty one." Raul said.

Jill turned to her side and looked at Sam with squinted eyes. He put his index finger up to his lips as he smiled and winked at her.

"Okay, we are here, just continue through the main gate, then turn left," Raul directed.

"But the bus stop is straight out front," Sam turned and said.

"Yes, but my office is to the left in the park office. There you will have a seat and rest while I talk to the park superintendent. We will then see where you will go."

Sam looked at Jill, raising his eyebrow as Jill widened her sealed lips. They continued in silence until they reached the main gate. It was busy with tourists getting off an arriving bus. The freelance guides leading donkeys and camels scurried around offering their services. Young boys and girls were bouncing from one tourist to another, offering bottled water and soft drinks. They gave warnings about how dangerous it was to enter Petra without bottled water, and that they had heard the park restaurant was closed. Every now and then, a park guard would have to yell at the young kids hustling for a dollar.

As they went through the gate, Raul said, "To your left, that is the office. Please go in the front door and be seated. I will see you in a few minutes." He continued to follow them to the front door. The guard who had been out in front of them the whole time opened the door and then stepped aside, allowing all three to enter first. Inside, Raul motioned for them to take a seat. He said something to the guard behind a desk and then went through a short swing gate to his office in the far back room.

Jill and Sam took their packs off and sat down. The guard behind the desk sat and stared at them. When Jill looked at him, he said something in Arabic and pointed to a refrigerator over against the wall. Jill nodded her head and said thank you. She went to the refrigerator and got two bottles of cold water. She sat down, handed one to Sam, and then opened her bottle and took a long slow drink.

As a bit of water dribbled down the corners of her mouth, she asked Sam, “So what do you think will happen to us now? I guess we broke the law staying in the park all night. God, who knows, we are liable to spend several years here in prison.”

“God, I hope not. I think they will just question us about what we were doing and why we did not go out the north gate. Let’s both say we just were disorientated and went the wrong way. Then we saw the lights of the town in the valley, so we decided to camp there and then go to that town when morning came.”

“Okay, good story, but …” Sam elbowed Jill and was looking at the guard behind the desk. He just sat there staring at them.

Sam held his bottle up toward the guard and said, “Thank you. It is good and cold. Do you speak English? Do you have anything to eat? You know, snacks or something?” The guard just looked at them for a moment and then down at the paper he was writing on.

Sam muttered, “Guess not.”

From the side of her mouth, Jill whispered, “Guess not what—no food or no English?” then added. “Why are these guys so curious about the old sheep herder, they may know each other but I do not think they talk to each other.”

“I don’t know but they do want to know what we have talked with him about, so let’s just be quite about that.”

Raul came back out after a boring twenty minutes and told them that they would have to wait there for the superintendent. He asked if they needed anything. Sam said that they would like to know why they were being held there and not allowed to go to Amman. Raul told them that the superintendent wanted to know why they chose not to do what they said they were going to do.

Sam stood up, walked over toward the desk, stopped at the little gate, and said, "Look, we apologize for what we have done. We simply got turned around a bit and went the wrong way. When we saw the lights, we even thought that might be the north gate. That's how confused we were. We caused no damage or harm where we slept. We were not going to walking around in the dark to find the back gate. No one would have been there anyway. Can't we just forget about this and leave?"

As Sam stood, still waiting patiently for a response, he stared at their two BlackBerrys laying on the desk.

The guard shrugged his shoulders and said, "That is fine with me, I understand your mistake, but you will need to tell that to my boss." Raul left the office and told them that the clerk would bring them some lunch when he went to get his. They were free to walk around, read, or do whatever they wanted—but they could not go to the door or the clerk might take it that they were trying to escape, and they would be shot. Raul then went out the door, leaving them with the clerk staring at them.

When the time came for the park to close, Raul came back. He informed them that the park superintendent had to leave town for the day and would be back in the morning to talk with them. Jill and Sam were shown into a back room. It was a small room used by the all-night guard who got stuck with the night shift. There were a couple of sleeping mats on the floor, a couple of chairs, and a table. It had a small bathroom and kitchen area. There was one window at the back of the room, overlaid with bars.

Raul motioned them to go into the room and said, "This is where you will spend the night. It is much safer than sleeping out on the mountain with the jackals. It will also save you a hotel cost. Is there anything I can get for you?"

"Yes, I think there is. Can we use a telephone to call our home? Can we have an explanation as to why we are being held against our will?" Sam demanded.

The guard motioned for them to go into the room again and said, "The office is closed, so you cannot use the phones. The batteries in your phones are dead. I am keeping you here as ordered by the park superintendent and giving you a good place to spend the night, for free. You committed a crime against the park rules; therefore, you must explain what you did to the proper person."

"What will happen if we do not go into the room to spend the night?" Sam asked Raul as he stood back looking into the room.

Raul smiled a bit and stared straight into Sam's eyes. "That is not a good question to ask at this time. You are in my country now. You will be smart to simply do as I tell or ask of you, because you are going to do what I say no matter what."

"Can we talk with the field archeologist who is working here?" Jill quickly asked as she gathered up her backpack.

"Yes, of course, you may speak with her if she comes to see you after the superintendent speaks with you. So, tell me, what did the old man tell you? Why did you go see him?"

Jill smirked at Raul, grabbed Sam by the shirtsleeve, and said, "May we please have our cell phones so that we can at least listen to music? We have charge cords to plug them and charge them."

"Ah yes, in the morning you may have them. They will be charged through the night, please remind me in the morning."

"Come on, Sam, we are not getting anywhere with this guy. We will have to wait until tomorrow to get any answers."

Raul called for the night clerk to come over to the door. As he explained some things to him in Arabic, Sam, and Jill dragged their backpacks into the room. He then told them to have a good night's sleep, and that if they needed anything, to just knock on their door and the night guard would come to them. Leaving the door open, Raul left for the night, and the clerk returned to his desk and soccer game on the office TV.

Jill burrowed in her pack for something to eat while Sam got their sleeping bag and bathroom kit out. They ate what fruit, cheese, and crackers they had left.

Jill did not feel like eating much, so she mostly sat and watched Sam eat the leftovers. She took a long drink from a Jordanian Pepsi and then said, "Every time we have had tea here, we have been visited by

Youssef—so we must have tea. Fix me a glass of tea, and maybe he will come and help us out."

Sam picked up his little camp boiler and poured Jill a glass. Sam raised his little glass of tea and said, "Here's to you, Youssef. We need your help and hope to see you soon." Jill raised her glass also in a salute to Youssef along with Sam. They sat and ate quietly, and then they talked for quite some time about who they should call in the morning and if they should try to call the U.S. Embassy. They both agreed that they would say nothing about seeing the ghosts but would only say that they did talk with an old herdsman. They agreed to stick with their story that they became a bit lost and were going to the town they could see. They finally cuddled up together in their shared sleeping bag.

As she rested her head between Sam's bicep and chest, Jill began to cry softly. "I kind of wish we had not met the ghosts. We could have been just like all the other tourists and said, *oh, I smell something funny* and then just went on our normal little tour. But no, we had to explore farther. I wish my mom were here."

Sam pulled her closer and said, "Jill, we will get things worked out in the morning. You are never going to be just like the everyday person. You and I could be at home now or on the boring beach somewhere, but we just could not live that boring life. Do not let this worry you. Just think about how you are going to explain it to your dad. Man, I can see him just laughing his head off. Now get some sleep."

Early in the morning, they were awakened by the loudspeaker of the prayer tower calling everyone to the first prayer of the day. They got up slowly, took a cold shower, and fixed some hot oatmeal for breakfast. After a couple hours of waiting, the door opened, and the guard asked them to come out and meet with the park superintendent. He led them to a closed office door, knocked on it, and listened for a response. A voice came from inside the office telling them to enter. The guard opened the door and motioned for Sam and Jill to enter. As they entered, the guard pointed to a couple of chairs for them to sit in. The superintendent sat behind his desk with his arms folded on his stomach.

After a few moments of silence Jill asked, “Why are we being held here?”

The man behind the desk looked at Jill but said nothing. He then looked over at Sam and said, “What do you have to say for yourself?”

“I have to say that we are sorry for camping out in the park. We broke no other violation and caused no harm to anyone or anything. We ask to be freed now so that we can go to Amman, visit the museum, and then return to our home,” Sam said as he sat on the edge of the chair holding Jill’s hand.

The man looked directly at Sam. Rubbing his chin with his hand, he looked over at the wall, at the ceiling, and then back to Sam. Finally, he asked, “What did the old man Youssef tell you? Why did you go there to see him?

Sam looked at Jill and then down at the floor, shaking his head a bit. Looking back at the superintendent, he said, “We met him at Turkmaniyeh Tomb; he offered to give us a tour of some of the sites around the park and tell us about the history from a Nabataeans point of view. He seemed to be a kind man and …”

The superintendent interrupted, saying, “Yes, I know, no money, so what did he tell you about the history and what special things did he tell you about Petra and the Nabataeans?”

“He told us about the history of the people here before the time of Christ and what happened after Christ and how the Roman Empire took over Petra. From that time on, the city declined in growth and soon became abandoned as a trade center, but that the Nabataeans still existed today.”

“A trade center for what, did he say?”

“He said they traded lots of things here, such as clothes, fabrics, foods …”

The superintendent interrupted again. “Okay, okay, okay, just say the word *spice* and tell me about that. What did he tell you about spices?”

Sam looked straight at the man and said, “He did not say much of anything about spice other than this valley was called the spice route until the Romans came. Other than that, I think it is just a tourist attraction and an archeological dream for some," Sam said and waited for the park official's response.

After a moment's silence, the superintendent said, "I did not ask you what you think."

"Gosh, what is so important about salt and pepper or whatever else they used on their goat meat a couple thousand years ago? May we speak with the archeologist from the U.S. who is working here, and can we use a phone here?" Jill blurted out.

"The office is closed. Today the park is closed. I have come here to talk with you on a day that I should not be here, but I am being kind to you."

Sam mildly burst out, "Closed today? It's Saturday. Why is it always closed? Can we not use a telephone on Saturday? Where are the local police, can we talk with them?"

The superintendent raised his hand, looked down at his desk, and shook his head. "Wait, wait, please wait. Please do not get so excited. First of all, today is the Sabbath, and there are no businesses open today. The archeologist and the excavation crew do not work on the Sabbath, as one must obey the rules and laws when in another country. And lastly, I am the official law of this park, and I am the chief of the police for this area. My name is Sharif." He then pointed to Raul, who had been sitting quietly in a nearby chair, and added, "Now, this man here who works for me has told me of your violation. That is, failing to exit the park at the closing time and failing to exit the park at the gate you so designated. Thirdly, for camping in the park, and of course I am sure you are married, or that would be another wrong here. Now, this man is either a liar and you have done no wrong, or you have done wrong and must account for what you have done. Which is it?"

Jill started to speak, and the man held his hand up to silence her. He then slowly sliced the air with his hand, gradually turning the palm up toward the ceiling and pointing toward Sam. Jill's fair-skinned face flushed to a burnt orange as she glared at the man and scooted back into her chair. She folded her arms and looked at the nearest wall.

Sam looked at her and gently squeezed her hand. He looked at both men and said, "We are not saying he is a liar. First of all, we failed to exit the park at closing and at the right location because we were a bit lost. Secondly, we camped because we were afraid to travel in the dark. Is it illegal to camp in the park? If so, there are others who are actually camping all the time up in the hills. The Bedouins do. Why don't you arrest them? So, did we actually commit a crime? If so, we apologize and will leave." Sam looked at the superintendent, at Raul, and then back at the superintendent. He put his hand on Jill's knee and pushed down a bit to stop the vibration.

Sharif sat forward with his arms out on the desk looking straight at Sam and replied that camping is illegal for tourists, not for the local people of Jordan. "Do you have anything to say about what the old man, Youssef, said to you?"

"No."

"Very well, we will continue this discussion tomorrow. Some food and water will be brought to you before late evening." The superintendent stood up and motioned for Raul to take them back to the holding room. Sam began asking for a phone call again, and if they could talk with the archeologist in the park but stopped talking when the door was opened. Just outside the door were two-armed guards. They did not just carry pistols strapped to their waists. They were carrying assault weapons. They stepped aside and let everyone out of the office. Sam and Jill went to their room.

As Sharif went out the door, he said, "Oh, by the way, the armed guards, they do work on the Sabbath. They are here for your protection."

Sam followed Sharif to the outside door and asked, "If a visitor like Youssef comes to visit us, will he be allowed in to see us, and will he be arrested?"

He raised his arm quickly, giving his hand a flip in the air. "Of course, yes, of course you may speak with anyone who comes to see you. Just tell me about it tomorrow. You are free to do what you wish—but you will not leave." He got in his car and drove off as Sam watched from the door. Raul spoke with the two soldiers and then left. One of them motioned for Sam and Jill to go to their room. The other one spoke to them in Arabic, pointed to the room refrigerator, and nodded his head toward it.

Sam went to it, got several bottles out for the day, then went straight the waiting chair and book in their new bedroom. To keep from getting bored, they read what guide and history tour books they had on the several sites they had visited. Amman was to be the last place to visit, and from there they were to fly out. If they were not home in two weeks, family members would become concerned. However, they were hoping Youssef would become concerned first. He was the closest friend they had within two thousand miles.

At about noon, they heard a knock on the outside door and heard voices talking. At least one was English. Soon a knock came on their door, and then it was unlocked and opened. The American archeologist stepped in. The door was left open, and the guards returned to their soccer game on the TV.

She put her hand out and said, "I am Dr. Mary Smithenburg. I don't think I properly introduced myself at our first meeting. I see you have yourself in a bit of a pickle. One thing I wish they would teach in various cultural anthropology classes is that laws and regulations should be learned before students are allowed into another country to do their amateur studies. However, what is done is done."

Sam folded his arms and asked, "Are you here to try to help us get out of this mess or to blame us for it? Really, we have done no harm. We were just caught camping because we did not want to walk the trail at night. Can you and will you help us get out of here?"

Dr. Smithenburg said, "I will speak with the superintendent in the morning and will tell him that you are just students in the U.S., and that I hope he will overlook your inexperience in the field of travel and meeting other cultures. I think he will forgive you and let you go soon—however, you must be patient and respect the methodology of these people. As students, you should conduct more research into the location you will be traveling in. Oh, by the way, do the two of you have some student identification for me to see?"

Jill, flushing burnt orange again, quickly turned and said, "I'll get our IDs." She went to their backpacks and dug around. As the doctor continued to preach, Sam could hear things hitting the wall and maybe the ceiling, but the doctor's eyes were just looking all around as she talked. Jill returned and handed their passport and ID wallets to Sam as she went to the outside door. She motioned to the two guards that she needed some fresh air and would be just outside the door. One guard nodded his head in acknowledgement and went back to watching soccer. Within a few minutes, Sam came out the door, followed by Dr. Smithenburg. As the doctor stepped outside, Sam asked her if she had ever heard of the ghost stories here.

The doctor said, "Oh, so this is where you were. I thought you might have gone to the store or something. Oh, but they are all closed. Oh, and I don't think that would be allowed anyway. Well, I will keep in touch with you. I will talk to the superintendent in the morning. Don't worry; it will be all right in a few days. Oh. Also, I would highly recommend that the two of you take my field-school class here next summer. That will give you some experience …"

Jill raised her hand and said, "Sorry, ma'am, but I have got to go to the bathroom. I beg your pardon." Jill went back inside and went to the guards. She pointed at the TV and motioned if she could watch. The guard waved his palm to the floor and then at the TV. She gladly took a seat on the floor and watched the game on Jordanian TV.

Sam loudly asked again, "The ghost stories—have you heard anything of that nature while you have worked here?"

The doctor said, "Oh yes, I have heard everything. And that is exactly what they are, stories, just stories like everywhere in the world. There are no grounds of evidence for such things or beings that can even begin to pass the scientific methodology test. Stick to the scholars and you will learn more …"

Sam raised his hand and said, "Excuse me, ma'am, but I have got to go, you know, *go*. We thank you for your help and hope to see you tomorrow." Sam quickly went inside and found Jill watching the TV. He joined her on the floor and said, "Shit, can you believe that lady is a doctor? Wow, how in the world do some get these great positions?"

"Double the shit for me," Jill said as they sat and watched the game and talked a bit. The day went by slowly until about five o'clock in the evening. Sam and Jill were locked back in their room so the guards could go home to eat. Shortly after they returned, another knock was heard from the outside door. The guards opened it and talked with a man in Arabic.

Jill said, "It's Youssef. Oh, thank God he is here."

Sam said. "Yes, but boy, we will have some answering to do tomorrow. Maybe he will stay, and that way we can all just sit down and talk this mess out."

Their door was unlocked, and Youssef stood holding out a fresh loaf of bread and a bottle of wine. "Hello again, my lost friends, I have brought you some dinner relieving the guards of that trouble."

"Thank God it is you. We are so glad to see you. Did they tell you what our crime was?" Jill rapidly said. Youssef stepped into their room and motioned for the three of them to sit and have dinner.

Sam said, "Yes, let's eat, I am starving. We have eaten all of our snacks and food. We have nothing for breakfast, so I hope the superintendent brings us some in the morning."

Youssef said, "He will. He will also want to have a long talk with you after breakfast."

"Yes, he wants to know all about you, and he wants us to tell him all that we have talked about and why we are meeting. What should we tell him?" Jill asked.

"Yes, why is he so interested in what you tell us? Do you know him?" Sam asks.

"Yes, of course, everyone here knows this man. He is the ruler of this town. In addition, this man knows me. He does not like for me to make a little money as a guide. He does not like anyone making money from this park except himself. I have spoken with him today. I came here to see if you were back at the hotel because you choose not to come and see me. At the entrance gate, I was told what had happened. So, I had the park guards call him and ask if I could bring you something to eat and drink. I was told that I could, but that I could stay no more than thirty minutes. So, let us eat." Sam grabbed the fresh bread, tore off the end, and grabbed a butter knife and the small round of white cheese.

As Sam spread cheese on his bread, Jill asked, "What do we have to do to get out of here and hopefully not pay a fine?"

Youssef took the bread from Sam and tore himself off a piece, picked up a slice of meat, and said to Jill, "Please, pour us a glass of wine, the glass you will find in my bag there." Jill watched the two of them eat for a second or two and then grabbed the bag. She opened it to retrieve the tea glasses and looked inside. As she reached inside, a great smile grew on her face.

She closed her eyes and said, "Oh my God, they are here." She quickly closed the bag and asked, "Do I let them out, or I mean, can they get out? It is Latif; I know for sure, I would know nutmeg from anything, as much as I like pumpkin pie." Youssef put his index finger to his mouth as he chewed his dinner and then pointed at the wine bottle. Jill opened the bag up again and took the three glasses out. After she poured them full of wine, she asked if she should leave the bag open. Sam asked Jill if she could see them in the bag, and then he looked at Youssef and asked how many came. Youssef said that only one had come with him, and that was a danger to him as well as Latif. He then told Jill to leave the bag open so when the time was right, Latif would appear and help them if the guards went home for the night. He told them to continue to watch TV with the guards if they allowed it. This would put them at rest, and they would not suspect anything. He told them to go to bed before the guards left, which would be about ten o'clock. With the lights out and being quiet, the guards would go home for the night.

After about thirty minutes, one of the guards came to the room where they were having dinner. He looked inside the doorway and told Youssef that it was time for him to leave. They all got up and walked to the outside door. Jill and Sam stood in the doorway and waved goodbye as Youssef walked up the street toward the town center. Sam and Jill asked to watch TV. One of the guards gave them the standard horizontal palms up motion of the hand to take a seat on the floor. As Jill sat on the floor and the two guards continued watching the TV, Sam went to their holding room and came back with the bottle of wine. He placed it on the receptionist's desk next to the BlackBerrys and offered it to the guards. One got up, bowed his head in thanks, and got two glasses from the small kitchen for him and his partner. Jill and Sam watched the soccer game on the TV as the two guards yelled and pointed at the screen and each other. At about nine o'clock, they got up, said goodnight to the two guards, and went to their room. Jill could not resist picking up the bag Youssef left. She looked in it and found another bag containing a brown powder and strong fragrance of nutmeg.

She started to put her hand into the bag when Sam quietly yelled, "Don't touch it. You might disrupt it or harm it. It might even do something to you. Who knows, it might turn you into a pumpkin, or maybe a ghost." He then grinned really big at her and said, "Jill, it's not time to let the ghost out of the bag."

Jill looked at Sam with a smile completely across her face and said, "I just want to touch it. I am simply curious about how they are just a little bag of spice at times and then turn into a ghost human."

"That is where I come in. You found them, now give me time, and I will tell you all about them … I think." Sam said.

Jill carried the bag over to Sam and held it out. "Here, you test it."

Sam took the bag, looked in it for a moment, and then put it in the corner of the room where he could see it from their sleeping bag and said, "Let's just let a sleeping ghost sleep and wait until he comes out on his own and see what he has to offer in help."

"I bet if it were another ghost in the bag, you would not be saying that. It would be me."

Sam stepped away from the ghost bag and got ready for bed, keeping silent from Jill's last comment. Jill followed, turned the lights out, and went to bed. They both tried to stay awake and wait for Latif to appear but fell asleep with just the aroma of nutmeg in the air.

About midnight, they were awakened by a quiet voice calling their names. They did not move other than grabbing each other and whispering.

"Jill, did you hear that?'

"Yes," she whispered back. Sam always slept with his flashlight next to him. He found it with his hand, pointed it toward the ceiling so the light would reflect into the whole room, and then turned it on. Just as the light illuminated the room, they both jumped into each other's arms. They stared toward the bag and found Latif standing there looking at them.

He put his hand out with his palm facing them motioning for them to be calm. He whispered, "Pack your things and I will lead you to where we are to meet the son of Youssef at the edge of the town."

They quickly began gathering up their things and packed their two big field packs while glancing at Latif. When they were about finished, Latif went to the door. He put his hand on the doorknob and held it there for a few seconds, and a click was heard from within the door. He took his hand away from the knob, allowing Sam to open the door, and then the three of them crept into the main room. The guards were not there. They had gone home for the night as Youssef had said they would, but the two phones were still on the desk. Jill darted back to the desk and grabbed them. Latif and Sam went to window and looked out into the street.

Sam said, "It looks all clear, let's go."

Latif put his hand up, signaling them to wait. He put his hand on the doorknob as he had the room door, and a click was heard. He whispered, "It must have a key from the inside also."

Jill asked, "How did you do that?"

Latif said, "the locks are electronic, and I can just make it open."

Sam then opened the door slowly and they quietly went out into the street, hugging the shadows. Latif grabbed Jill's hand and tried to lead her into a dark alleyway. Jill froze up at first as a shiver tingled through her entire body. As she shuddered, goose bumps streamed from the back of her neck to her feet. She grabbed Sam's hand then followed Latif. They walked hunched over a bit and very quickly, without a word. Soon they came to a small white pickup truck. Latif told them to get into the back and lay down. Latif told Sam to put Youssef's bag in the passenger's side of the truck. When he put the bag on the seat, he could see only a glimpse of a middle-aged man in the driver's seat. He let Latif float into the seat next to his bag and then closed the door and climbed in the back next to Jill. The motor started and they drove off.

Jill said, "Wow, that was easy. Now we are free again. Maybe we can go to Amman tomorrow."

Sam said, "Maybe that was too easy. Maybe we should have just stayed there and hopefully got out legally. Now we are escaped criminals. But anyway—did you feel that tingle? Holy smokes, I felt it all the way through your body into mine. I just hope it was your tingle and not his."

"Sam, think about it, it was his. It just came through me. Do you feel that tingle with just me?"

"Well, yeah, sometimes. You know when we huchech…"

"Sam! Get serious. We still are in deep shit."

"I am serious. It was your tingle. You see, electrons push one another like water in a pipe, so I think they have something to do with electron flow. His just pushed yours into me and …"

"Okay, just peep over the side and see if we can figure out where we are going. There is one thing for sure—I bet we won't be able to hang around Youssef's place exceedingly long, because that will be the first place the officials go."

After about an hour's drive, they came into Quweira. They could tell by seeing the signs painted on some of the building walls. The truck made a few turns here and there and ended on the other side of the town with few houses. The truck stopped under the carport of a small two-story house. As soon as they stopped, the driver got out, came to the back of the truck, and lowered the tailgate. In a broken English he said, "My name is Yaman, Youssef is my father. You are to stay the night here in this house. He will come to see you in the morning. Get your things and follow me."

As they went by the front of the truck, Jill looked in the cab. Latif was not in there, but the bag was. Yaman unlocked the door and quickly showed them where the light switches and the bathroom were. He then led them to a bedroom with bed mats on the floor. They went back to the main room, where Yaman went straight to the door.

"I must leave now. Use the lights for only a few minutes, then settle down, and get some rest. I must lock the door and do not have a key for you. You will be fine. Goodnight."

Jill stepped forward a bit and said, "We thank you for your help and hope we have not become a big problem for you and your family, and we are in debt to you."

As he went out the door, he said, "You are welcome. You do not owe me anything." He quickly got in the truck and drove off. Jill and Sam scouted the house out and found that they were alone in a new house. It had three small bedrooms upstairs, a kitchen, and dining area downstairs along with a bathroom and living room. They gathered their gear, went to a bedroom, and settled down to rest and hopefully to sleep.

Chapter 3
Quweira

The morning sunlight filled the room early through the curtianless windows and soon woke Jill. She shook Sam and told him to wake up. The first thing they did was to go to every window upstairs and look outside to see if any activity was going on, such as a park guard's truck driving around. Nothing much seemed to be going on, other than a few cars and trucks going up and down the road outside. A few people could be seen casually walking the streets, since Sunday in this part of the world was like a Monday in other parts, and people were just getting back to work and school. Sam and Jill left all their gear upstairs in case someone came and looked in the windows. They even kept most things packed in case they had to make another quick exit. After they washed up a bit and brushed their teeth, they gathered enough stuff to go downstairs to fix a pot of tea and maybe find something to eat. They had started preparing some hot water for tea and oatmeal when a soft knock came at the door. Then a key could be heard unlocking it. It was Youssef, two other men, and a lady. Youssef introduced them as his wife and two sons. One was Yaman, who had brought them to Quweira in the dark. The other one's name was Asad. Youssef said he had a third son named Imad who was busy at work.

Jill asked, "Where is Latif?"

"He is fine. He is with his friends at Al-Barid, where they stay.

Jill asked, "And just where is that?"

Youssef took a deep breath and said, "It is what is also called Little Petra, about ten kilometers to the north of here. Yaman, my son, will take you there now. I will come to see you late today or maybe tomorrow. Nevertheless, I must stay here and wait for the guards, who, I am sure, will come here looking for you. I brought you some things to eat. Here, take this sack of food and go."

Youssef's wife and oldest son remained quiet the whole time. When Youssef was finished talking, he motioned for them to go out the door where Yaman stood waiting by a little white truck with tall sideboards painted with colored flowers. They grabbed their oatmeal cups and began stuffing spoonful's into their mouths as they scurried up the stairs to grab their packs. When they returned downstairs, everyone was gone except Yaman. He motioned with the spinning of his hand for them to hurry up. Jill grabbed Sam's bowl and took them to the sink to wash them out as Sam took the two packs to the truck waiting outside. Jill came outside with the two bowls, two full cups of tea, and the cook pan. Sam was already in the back of the truck waiting and took the warm spilling tea glasses from Jill as she climbed in. Yaman helped Jill into the back and told them to lie down.

As he went to the front of the truck, he looked through the side rails and asked, "You have everything, you leave nothing, yeah?"

"Yes, we have everything, we are ready." Sam replied.

Yaman got in the truck and drove off. Sam and Jill laid low and tried taking a few sips of tea but managed only a couple before most was spilled.

"I hate to tell you this, love—but we don't have everything. I forgot the international electrical adapter," Jill said softly.

"No, you didn't, it's in my pack, we did not use it here," Sam assured her.

"Uh—yes we did, or I did. I plugged it in to charge the camera back at the ole jail and forgot it in the mad rush we left in."

"Shit, and that was our only one. Well, maybe we can find another one. For now, let's just find a way to get out of this mess." They had to interlock their fingers and put their hands behind their heads to keep from banging their heads on the floor of the truck bed. They only had to do this when another car came up behind them, because resting on the packs allowed them to be seen over the tailgate. After about an hour's drive of who knows which way, the truck slowed down. Sam raised his head and looked through the slats in the wooden sidewalls.

"Man, it looks like we are back in old Petra. But this is a much smaller valley. I can see a few carved facades and some that just look like caves."

The truck continued slowly. Soon they could remain flat on their backs and see mountain walls on both sides of the truck closing in on them. The truck stopped beside a great-carved structure just a few feet away. Sam and Jill sat up as they heard the door open. Yaman came to the back and lowered the tailgate. They grabbed their bags and climbed out. Yaman walked over and pointed at a small cave-like opening in the mountain wall.

"Here you stay. You wait for my father today, maybe tomorrow." He then pointed over at a small oasis and said, "There you find water. Now I go."

Jill asked Yaman as he walked back to the truck, "The water, can we drink?"

Yaman replied as he got in the truck, "Of course you drink, yes, of course." Then he drove off.

Jill looked at Sam sadly and said, "Dear God, what have we gotten ourselves into. Where in the hell are, we, and what's next?"

Sam took the two packs over to the carved cave-like opening and looked inside. From the smoke stains at the top of the doorway, it was obvious people had stayed in the little carved house before. It was dark inside, but the back wall could be seen easily from what light came in through the only door. He put the packs against the wall inside and then told Jill. "Come on, let's have a look around. We can climb the ridge over there above the oasis and see a lot. We must be about ten miles from Quweira, like Youssef said, about ten kilometers. From the way this narrow valley looks, it must be a dead end. From up there, we will be able to tell."

They got their small backpacks and some water and climbed the steep valley. Before starting their climb, they crossed the valley floor and entered the oasis. The spring came from a large crack that went back a ways into the hillside.

After they climbed a bit, while taking a break, Sam looked down on the oasis and said, "You know something … I think we will move over to this side of the valley. We can camp right down there in that oasis, and there is that carved structure right there with water within ten feet of where we can put the tent. If a storm comes, we will have that structure for shelter. If we see some stranger coming, we have this hill we can climb and hide easily. What do you think?"

Jill looked around and said, "That would be a much better place to stay; that way, we have water nearby and those bushes for a bathroom and some firewood."

Sam chuckled as he got up to walk and said, "Well I don't need the bushes out here, so I let you gather the firewood."

Sam had to hurry on up the hill to avoid a flying stick and rock. When they reached the top, they sat on the highest point and looked around. They could see to the end of the valley, which was just a mile or so away.

"Wow, I see why they call this valley the Little Petra. It is just smaller in width. There are a lot of carved structures you can see from here to the end of the valley where we drove in."

"Are there any really important ones according to the book?" Sam asked.

"No, not really, this place served as a large storage area for the main Petra. It was also one of the few entrances. The caravans coming from Egypt and Gaza had to enter Petra through here first. As narrow as it is, there was no way anyone could sneak in or out," Jill instructed.

Sam stood up, walked around looking up and down the valley wall, and said, "Well we are going to have to figure out a way to sneak out. Isn't this place also known for having many more tombs than the main city? It seems like I remember the first day we were in Petra, a man in the little museum saying there was a big burial town to the north of Petra, and I am quite sure—we—are—to—the—north."

"Yes, my dear, according to the map in this book, if we are in Al-Barid, we are just a couple miles from the north gate of Petra, by the way the crow flies anyway. We went south last night and then turned to the north to go around the mountain. So, if all else fails, we can walk back into Petra and turn ourselves in as captured by the ghosts. I think that is what Raul and his boss are looking for. They wanted to know everything

Youssef told us. So, they either know about the ghosts and want them, or are suspicious about them and are trying to get it out of Youssef."

"Yeah, you are probably right. Or maybe it has something to do with the hidden treasure that people are looking for here. Well, let's go back down, move our things, set up camp, and then walk down the valley and see if we can sniff out a ghost. We know they are here," Sam said as he reached for Jill's hand.

After a steep downhill descent, they scouted out a camping site and set up their tent, gathered some firewood, and filled their empty bottles with water to boil later just to be safe. After camp was ready, they went for a walk, hugging the valley wall and going from structure to structure.

Sure enough, they could smell various fragrances of spices. They did not talk about it aloud, because they still had not learned if the ghosts could hear them while in the dormant state. They looked around, hoping to find a pile of spice, but found none. At one large structure, Sam got down on his hands and knees and sniffed around inside the building hoping to find a ghost in the sand. After about a mile of walking and searching, they turned around and went back up the other valley wall of tombs and spice storage rooms.

Finally, Jill said, "What we need to do is just sit around one of these buildings and wait for them to come out."

"Yeah, that's a good idea, but maybe they only come out at night, because we really have not seen one in the broad day light hours. Also, what if they are not friendly?" Sam asked.

"Oh crap, I don't know, but let's not talk about that. I mean, the three we met were not mean. Really, they were nice. I think they—want something from us."

They walked quietly for a while longer and then Sam said, "You know, if they were a real big danger, there would be rumors flying all over Petra, Jordan, and in worldwide ghost stories. But I have never heard of ghost stories here, and when we asked people, they did not think much of the idea of ghosts being around or were kind of weird about it." They gave up on trying to find a ghost and went straight to camp. It was getting late in the day and would soon turn to darkness. They built a small fire in the structure that was built around the crack where the spring came out. That way, it could not be seen by anyone who came up the valley. They were not in any hurry to go to bed, so they boiled water to sterilize it and filled all their empty water carriers. Then they made some hot tea, hoping Youssef would arrive. Whatever they did, they did not take their eyes off the valley and its carved city. Occasionally, Jill would softly call out to the ghosts, asking them to show themselves. As it got dark, Sam and Jill put their headlamps on their heads and turned them on once in a while to look out through the darkness. The night out here, they learned, was a lot different from darkness back home or in any town or city, where lights were on all night. Soon they could not see across the valley floor or the opposite valley wall, which was only about 100 yards from where they sat, even after their eyes became adjusted with darkness. When they finished their last cup of tea, they put their sleeping bags beside the tent where they could stay outside, lean up against the structure's outside wall, and watch. They stayed up for a couple hours and finally drifted off to sleep with their backs taking the place of a couch back.

Sometime after midnight, Sam was awakened by a strange scent. He had thought about what the man had said about being awakened by the smell of spice. Over and over, he thought about this, which is somewhat like programming one's mind to be an inside alarm clock. Before turning on his headlamp, he gently shook Jill.

She whispered, "Did you hear something, or did you see something?

"Shhhh," he whispered. "Look out across the spring and on the count of three, turn your headlamp on and look straight out, then slowly turn your light to your left. Ready—one—two—three." Both lights came on, illuminating an area directly in front of them. Before they could scan their light to the side, they became frozen other than their heart rate. Their pulse went from a sleep rate of about 50 beats per minute to about 70 beats per minute with a startled awaking, then to at least 100 BPM within a few seconds. A small crowd of glittery human figures stood in front of them. Jill grabbed Sam; Sam grabbed a big stick he had put beside him. If they could have, they would have pushed the mountain back, but instead they just pushed sand up in front of their feet.

"Shit," Sam gasped,

"Oh my God," Jill cried.

The ghosts just stood still, except for a couple in the back who backed up and to the side to get out of the light beam.

Without thinking, Jill called out, "Latif, Saud, Haifa, are you there? Oh God, please be there." There was no answer.

Sam asked, "What do you want? We mean no harm to you. We are friends of Youssef."

They began to move around within their group, and whispers could be heard. There were really not that many, it just seemed like a bunch—or really, they just blended too much to be able to count. They all stood together just staring at Jill and Sam and seemed to be closing in on them like a pack of hunting dogs crowding around a helpless fox with no place to run.

Finally, a ghost to one side, just out of the light beam, said, "You came here, and a couple of us who were active or visible heard you calling us. So, we came here to see you. We know …"

The ghost, being out of sight, startled Sam, causing him to throw the bat-sized stick in his hand toward the voice. His elbow hit the wall he was leaning against, causing him to fall short of his target. The large stick hit the water just before the ghost. A great splash could be heard, and water spray could be seen hitting the ghost. As their headlamps illuminated the ghost figure, they saw him somewhat crumple under the water's weight. He kind of limped off into the darkness until they lost sight of him. As Sam and Jill turned toward the others, they could see several of them vanish into the dark canyon. The ones that were the closest took a few steps backward.

The one in the front said, "We are not here for trouble and came to see who was here and why you were asking for us. For you to ask for us to show ourselves, we realized that you must have met some of our friends. The three you call for are not here now, but we have sent word for them. They will be here in the morning. We will now leave."

With that, the ghost turned around and moved off into the darkness.

Jill gasped. "My God, they are so hard to see in the dark. They don't even show up well with our lights on them."

Sam said, "That is because they have a tint to them. The slight color gives them a camouflage like color gives most of the animals in the depth of the ocean. In the deepest places on earth, one would not think the animals have a color or that it would aid them to not be seen and eaten. But reds, blues, and other bright colors blend well with the darkness." He then crawled out of the sleeping bag and walked over to where the ghosts had been standing. He looked around with each step he took, looking to see if there were any lurking around. He saw none. He then went over to where the stick had hit the small stream. As he reached down to pick it up at the edge of the spring, he caught a whiff of evergreen trees. There were no evergreen trees within a hundred miles that he knew of.

"Hey Jill, come here for a sec. But come the same way I did."

"What is it? Do you see one? Are they waiting out there for us?" Jill asked.

"No, just come over here and listen. I think I can hear them talking off at a distance."

Jill crawled out of the sleeping bag and then wrapped it around her shoulders and walked across the spring and up behind Sam, who was kneeling down. Jill slowly spun her head back and forth, looking in every direction as she squatted down beside Sam. She put her hand on his knee to steady herself while scanning the darkness beyond her light beam.

She whispered, "Sam, I do not hear anything except you breathing and my heart pounding. But I smell something. And it is not the smell of a bakery. It is more like a campfire. It reminds me of a church camp I went to in Glorieta, New Mexico, when we had campfire meetings. Do you still hear them?"

"No, I don't. I did not hear them at all. I did not want to tell you I smelled something. I also smell something. It is the fragrance of Frankincense or myrrh. They both come from evergreen resin. It is right here where that ghost was standing who kind of limped off."

"I wonder if something was wrong with him, or if you hit him. God, I hope not," Jill said, looking at Sam up close. "Maybe he was just old. Let's get back over by the tent and watch from there. Get the stick, and let's find one for me."

Sam grabbed Jill by the hand and walked back to the tent. While Jill continued to scan the valley with her lamp beam, Sam searched the ground. They put a couple sticks and anything else they could find to use as a throwing weapon in a stack on both sides of their joined sleeping bags. Jill put the little camp stove by her side and heated up some water for tea. Just after getting the stove lit, she spilled the pan of water on the flames, putting the fire out. After drying it off and getting it going again, she turned a wide-lid litter bottle over, spilling most of the water. Sam got the tea and cups out of the packs and took over the job as Jill pushed back on the wall and looked out across the spring.

"Maybe we should move into the building for the night," she said.

Sam put his hand on her lower leg, trying to hold it still, and said, "No, I don't think that would be a good idea because we could not see out very well. I also think it would be better to be able to run rather than be locked in a cave as they were. All we can do is sit here until morning and then figure out how to get the hell out of here."

"I hope Youssef or the other three show up early. We need an explanation," Jill said.

"Yes, we do, but I don't think they want to harm us or kill us, or they would have. We will find out in the morning. But, for now, try to get some sleep while I watch. We may need the rest if we have to walk out of here," Sam said to Jill as she snuggled close to him, still clutching a stick.

Jill leaned on Sam and finally fell asleep with her headlamp still on. Sam turned her lamp off then sipped on his fourth cup of tea, sitting in silence and darkness listening and trying to watch. They needed to save their lamp batteries just in case they found themselves trying to run out of the valley in the dark. After a couple hours of vigil watch, Sam's anxiety wore off and he fell asleep leaning on Jill.

They were awakened by the sunlight after it was already on the valley floor. They did not have to get dressed, so they began gathering their things up and packed their packs in case they had to make a run for the hill or the way they drove in here. That was not the best idea, because back down the valley was where the ghosts were. Before packing or doing anything, Sam first went over to where he threw the stick. He got down on his hands and knees and made like a hound dog.

"It's gone. The fragrance of the spices is gone. There was no breeze last night to blow it away. The night air is heavier than the day air, so it would really push the scent to the ground. But it is gone."

"Well, Sam, it was not as strong as we smelled the first day or what we found yesterday. It was kind of weak, so maybe it just drifted off during the night."

"Or it was carried off," Sam said as he looked around. "They don't leave footprints. They seem to be somewhat solid when they want to be. They return to spice when they sleep or hibernate or whatever they do when they are not active or alive," Sam said as he analyzed their experience.

"Sam let's go. Let's get out of here. We can walk down the road until we get back to the town or maybe run onto Youssef."

"Fine with me, let's go. Grab your stick and let's get the hell out of here."

Within a few minutes, they were walking down the faint road. They ate as they walked, both carrying big sticks. When they reached the end of the small valley, they turned south on the better road and headed for Quweira. After about thirty minutes, they spotted a small house with a lot of stone fencing around it. There were a couple trucks there, but they were too far off to tell if it was an abandoned place or occupied. They headed straight for it. As they got closer, they could see sheep and goats in a couple of the pens. When they were almost there, they could see a moving dust cloud down the road. They just stayed on the road, hoping it was Youssef or someone they could get a ride with. As the truck got near them, it slowed down, and they could tell it was going to stop even though they made no signal for a ride.

Soon they could tell it was Youssef. He did not even pull off the road. He just rolled the passenger-side window down and said. "Quickly put your things in the back and then you can get in up here."

They put their packs over the tailgate, chucked their sticks in, and then went up front and got in the cab. Jill got in first and sat in the middle. She had to pick up Youssef's bag and another and put them in her lap to make room for Sam to get in.

She then quickly asked, "Why are you in such a hurry? Is something wrong? We are so glad to see you. We were getting worried, and we saw a bunch of ghosts last night!"

Sam leaned forward to look around Jill and said, "We were headed back to Quweira. We were planning on catching a cab there or bus and going to Amman. We have seen the ghosts and now need to get back home."

As he talked to Youssef, Jill pointed at the water bottle Sam was resting on his knee. Sam unscrewed the lid and handed it to Jill. As she took it from Sam, a small bit splashed out in her lap when they hit a pothole.

"Ah, please be careful with the water so you don't spill in my truck. We take water very sacred here in the desert," Youssef said as he took the bottle from Jill and then looked down at the bags in her lap. "I am in a bit of a hurry because the park superintendent sent four guards to my town asking questions about you. They came to my house questioning me and my family. They are going from house to house asking everyone if they have seen you. I believe they will come this far to that house back there. I will go there after I let you off at the place my son left you. I will see if the man who lives there saw you."

Sam asked, "Can you not just take us to your town and find us a cab or bus to Amman? Then we will be out of your troubles, and we can get out of ours."

Youssef looked down at the bags again and asked her to just put them on the floor. "No, you cannot go there today. You would get caught, and I also would get into trouble for helping you. This man at the park does not like me and would do anything to put me in prison. I will come back today if I can get away without leading the guards here to you. The ghosts you have seen will not bother you. I have brought the three you know to keep you company. They will watch for you and help you to hide if the guards come out here. Just make sure they are in their bag near you when you sleep so that they can alert you if someone comes in the night."

"Can they do that while they are in the bag?" Sam asks

"Yes, of course, they can do anything there."

"Were they at your house last night with you?" Sam asked as he rolled his eyes to the side, getting Jill's attention.

"Yes, they have been with me the whole time. They are with us now," he said as he pointed at the two bags near Jill's feet.

"We really need to get to Amman, but I guess we can stay one more night, and I will feel much better with Saud and the other two. What do you think, Sam?"

"I think we will be fine. We have survived two nights with ghosts, so I guess we can make it through another."

They arrived at the small-carved structure. Sam got the packs out of the truck as Jill gathered up the ghost bags, as Youssef called them. Youssef took one of the bags from Jill, took it into the building, and said, "Just put the bag on the ground in here and open the top a little, and then just leave them. They will appear just after dark. The bag you are holding has another bag in it. Take it out and put both of them here besides this one."

As Jill put the two bags down, she knelt over one, looked into it, and said, "This is Saud and this one is Latif, right?"

"Yes, you are correct. Okay, very well. I have a small box with some food for your dinner. I must go now and get back to my house before the guard starts looking for me." He turned and went back to his truck as Jill and Sam followed.

Sam went to the truck and got the small box out, opened it, and then asked, "So what is for breakfast? There is only enough here for the night."

Youssef got in and started the truck. "I will bring you breakfast." He then drove off as Sam and Jill stood still, watching him leave. After a few moments, Sam grabbed the two packs and walked over to the spring area rather than where Youssef intended them to camp.

"Let's sleep by the cave building again where we slept last night." Jill followed carrying the two heavy walking sticks. After they dropped everything, Sam began setting up camp. Jill made a trip down to the grove of small trees to gather more firewood. She informed Sam that she was going to build a fire just across the spring and try to keep it burning throughout the night.

After they got settled in for their third ghostly night, they climbed up the hill just behind their camp, watched the valley, and talked.

Jill said. "Maybe the fire will keep them away. I think I have seen enough of ghosts."

"I know it might be a bit unsettling, but I have just got to sit down with them again and talk to them and try and figure out just what they really are. We are not here in just a ghost story. We are here with real human—spirits, or is it the soul of a lost person? Well really, I think they are one and the same. After all the soul must exist and I think, we may have met one—outside of the departed body.

"Well, I agree with you somewhat, I also have got to get a couple answers. I don't think the fire will keep them at bay, but I think the water will," Sam said as he watched the facade door where Youssef had dropped them off.

"Why do you say that?" Jill asked.

Sam suddenly pointed across the valley and said, "Look, there is one coming out of the building. Let's just watch for a minute or two. You know, last night when I threw the stick and splashed water on that one."

"Yes," Jill said in a whisper.

"Well, what I think is that water shuts down their electrical impulse and flow. I think I wounded that one, and he lost some of his spice. That is what we smelled. Later that night I believe he or the others came and picked up what might have fallen off of him after it was deactivated by the water. They are afraid of water. Now we are going to test them," Sam said as he got up.

Jill grabbed his hand to be pulled up, agreed, and then said, "Whatever you do, just don't make them mad. How are you going to test them?"

As they descended toward camp, he told her what he was going to do. When they reached camp, Sam hollered out to the now three ghosts as they stood just outside the old storage facade. He and Jill continued to walk to them. They stopped a few yards from the ghosts and said hello and how it was good to see them again, as well as in the daylight.

Saud asked, "Were we supposed to meet you here at this facade and help you as you waited for the way to be clear for you?"

"Yes, I believe you were, is that what Youssef told you?" Jill asked.

"Yes, that is what we were told. So here we are. I guess we will sleep here, or really, I think we will stay active through the night so you can rest without worrying about the other ghosts. But let me assure you, they will not bother you," Saud said.

"Well, come on over to our little camp and tell us about the life of ghosts, or as you prefer, lost souls," Sam said as he motioned with his hand toward their camp. He and Jill then started walking that way, but the three did not follow.

Jill tugged Sam's hand, turned, and said, "Come, please. Tell us about your life, and we will do the same or whatever you want."

The three looked at each other and then, with a bit of hesitance, followed Jill as she turned and headed for the camp. When they all got to the spring, the ghosts stopped and sat about ten feet from the spring.

Sam and Jill stopped just after crossing the spring near their tent and then looked back to see the ghosts had settled near some rocks.

Jill said to them, "You are welcome to come into our little camp here, don't be shy."

"We are fine here resting among these rocks. There are some you can join us on."

"Ooookay," Jill said as she and Sam went back across the spring and then asked the ghosts to tell about the past and how they became trapped here as lost souls. Sam sat near enough to the water's edge of the little spring, playing in the water with his walking stick.

As he watched Sam making splashes in the stream, Saud told them how the people of Petra lived somewhat at peace with themselves and with the trade caravans that passed through their lands. Petra was a rest stop that provided food, shelter, and most important, water to all who passed through. As the Roman Empire grew in size and strength, the Romans also began forming a new religion, which gave them encouragement to further influence other nations and cultures into their way of life. The Romans became very attracted to the city and its strategic point in the trade route from the west area of Egypt and the east as far as trade was going. The Romans made it a part of their empire. The king's temple was taken over by the Roman commander and his captains. The king's daughter was taken captive as well as many others. The Nabataeans were given the choice to work as slaves under the Romans or become entombed in the tombs or spice storage rooms that had been carved into the mountains as you see today. Some Nabataeans managed to escape to the desert and to this day their human offspring exist and live there as Bedouins.

"The ghosts you see today were the ones captured and imprisoned in the mountain structures. The doors were sealed solid with marble stones and plaster making them airtight. We are soon all suffocated, as we lay weak and dying in the only beds there. These were beds of spice. Our souls were trapped within these walls for years, not able to go through the stone and on to the otherworld. Once we were able to gradually flow out through the stone and cracks, we were free. But we were beyond the time a soul has to seek the path to the otherworld. During the sixth century, Petra fell under the Byzantine Empire. The Christian religion was beginning to form, and the region was struck with a great series of earthquakes. These great earthquakes erupted, cracking the mountains, and spilling great quantities of stored water into the sealed chambers. There were many water cisterns found naturally or carved out all over the cliff tops to provide the city water throughout the year. This catastrophe left the city without water-gathering sources. Most of the Nabataean descendants had escaped or simply died out, and the Romans did not care to rebuild the great water ducts and cisterns. Petra soon became abandoned for good, with only a few Bedouins passing through once in a while.

"The only other remaining ones were us—the lost souls. While some souls were washed out through cracks left by the first earthquakes, others remained for years trapped. Over time, they seeped through the sandstone mountain walls and gathered in the sand. The grains of spice stuck, allowing the power of the soul to stick together and remain as the person. We have roamed these valleys now for hundreds of years, hiding here and there in the carved buildings, in the hills and oases. We have only been discovered by a few of our people's ancestors, and they have understood our situation and given us their honor of not disclosing us. We do not eat food or drink water. We hold a shape or figure of our once human person, but we are not that complete person. We can do many things as that person, such as move and talk. Yet we cannot cry, and we cannot die. We cannot enter into the afterlife as most all other humans do. We have learned that we once lived as you do today. Our body dies while our soul remains, which will never die …"

Sam interrupted, "Wait, two questions. Do you as ghosts still follow your old traditional religion, and would you mind if we took your pictures?"

Jill elbowed Sam as the ghosts looked at them and said, "Uh, don't worry about the second question, we don't have a camera."

"Dang, Jill, don't tell me you left it also."

"Okay, I won't, we have the camera, but it's not charged I used it to much at the castle and old church." Sam just sat with his head in his hands as his elbows rested on his knees.

Saud continued, "We do not participate in any religious practice, for over the years we have learned these are simply myths, so remarkably interesting yet simply myths. Humans almost require myths in order to survive as a social animal. But as true souls we learn and live by the truth that we are simply part of the cosmos of this one and only Universe. We would not want our pictures taken because they would only be used to expose us and torment us more than we already are. Also, we are grateful that our pictures can only show a slight cloud. The heat of fire does not bother us. We seem to be resistant to fire, but do not care to be in fire, and it is possible that lighting does not mix well with us. We require energy however a bolt of lightning is way too much. We are fairly certain that it actually scatters a soul into the air."

"Ooookay," Jill said. "Now tell me if there are any really neat facade structures or buildings around here that you could show us before it gets dark? This is really neat to sit here and listen to people who actually lived right here two thousand years ago. I mean, wow, I could learn from you people what really went on here and go back home and fix the books—and some teachers."

Latif said, "Let's take them to the Painted House. That is just about the only house that remains with original paintings in it. All of the other archeologists over the years have loved that place."

"Oh, yes, I have read of that place and seen pictures of it, and it is close by here, right?" Jill said.

"Yes, it is just around the bend in the valley there to the north," Saud said as he pointed in the direction that Jill and Sam had not gone yet.

Jill got up, grabbed her small pack and said, "Let's go.

Sam stayed cross-legged by the spring as he watched the ghosts simply rise up as if they were in the teleport room aboard the USS Enterprise.

As the ghosts started moving off, Haifa turned and looked down at Sam. She asked, "Are you not coming?"

"Well, I would rather not. I have seen the pictures in books."

"Well, just kick back and relax while I go learn something from these living time capsules," Jill said as she and the ghosts headed for the small narrow part of the canyon. "That is learning history from the people who made it. We shall return soon."

Sam waved at them and said he would be right here where he sat when they returned.

After the four of them had gone a short distance, Haifa stopped and said, "I will wait here with Sam. I have seen the smoky house hundreds of times, and it is a bit sad how over time things of one's life become decayed and how people just speed this up." She turned around and walked back to where Sam was sitting, and then stopped several feet away and watched Sam splash in the water.

Sam stopped hitting the water and looked at Haifa, and then back at his stick. He dropped the stick and turned around, facing her, as she continued toward him.

"Have a seat and we will carry on like we were. Do you like music?"

"Yes, I really like some music, others I do not like," Haifa, said with what appeared to be a smile.

This provoked Sam to ask, "So do you have emotions and feelings? What kind of music do you like? Probably nothing—modern!"

Haifa settled on a rock and said, "Yes, we have feelings and emotions just like when we were human. I like all kinds of music, some that is old, and I have heard only on instruments, and then there is that which comes from instruments like the little box you have in your shirt pocket. We are limited to listening to music we hear people playing because we have no material things."

"Ahhhh, my BlackBerry, would you like to hear what I listen to? I am very strange. I like music from Lady Gaga to The Rolling Stones to Bob Marley to classical guitar. I do not like 'crying in your beer music' or head-banger stuff," Sam said as he got up. He walked over to where she was and took a seat on the rock next to her.

"Yes, let me listen to what you like," Haifa said as she reached her hand out. Sam looked at her hand, her face, and then back at her hand. He wanted to reach out and touch it, but instead he reached into his pocket and retrieved his BlackBerry to show to Haifa. He said, "It does not work. The battery is dead, and we left the electrical adapter back in Petra. They were recharged at the park office but that is where we also left the charge cords. I have since ran the battery down."

Haifa continued to hold her hand out, turning her palm down, and then said, "Please let me feel it."

Sam held the BlackBerry out for Haifa to take, but she did not take it. She just covered Sam's hand with hers. A smile spread across Sam's face. Within a second, he closed his eyes. Music began to play at the same time a flow of tickle flowed through him. When he looked down at his hand, he could see light glowing in-between Haifa's fingers—the BlackBerry screen turned on.

"Now it is charged," Haifa said as "Crimson and Clover" played aloud.

Sam continued a warm smile as he rolled his eyes from the BlackBerry to meet Haifa's eyes looking at him. He said, "Yes, it's charged, but it was not before you touched it. So how did that happen? I want to know …"

"I know what you wanted to know … if I have a current if I was electric. Yes, I do. From the very beginning, I could sense you curiosity about us, but I do not sense harm from you as I do with the many others I have seen and heard over the years. We really do not need much water, and we do not like it because it can kind of shut me down, as you have discovered. And yes, I have emotion. I and the others here are like an electrical being. We are the soul of our once person, as you are today. I wish I were not."

Sam sat and thought for a moment in silence, and then he asked, with a small smile while looking at Haifa's face, "So, do you need the earphones?"

"No, I can't use earphones. We cannot hold or even grab things."

"I didn't think so, just thought I'd change the subject."

"Okay, you did. Now just let me hold my hand over yours as you hold the device. I will then see everything you have stored on it."

Sam held his hand out with the BlackBerry laying in it. As she put her hand over his, he had to close his eyes again as his face squinted. Haifa listened to the music as she stared out into the valley with the speaker off and read some of the menu.

"Ahhhh, yes, I like some of these."

"So, do you listen to them that way or just see them?"

"I see the list here in front of me like you see it. But I also see it in my mind. I hear what song I want to by selecting it. See, look at the screen, and watch me select this song. You see the song, "As Tears Go By," by The Rolling Stones. Now watch, I will highlight it, and voila, it plays. However, I hear it in my mind without headphones. Of course, I can hear it out loud just as you do also. I see us as maybe as organic artificial intelligence as smart people like you would call us." Haifa said as she removed her hand from hovering over Sam's.

He then rolled the little ball down to another song and pushed play. "Just Dance" played aloud. Sam asked, "So how does music make you feel? Do ghosts dance?"

Haifa slowly shook her head and laughed. "I am two thousand years old, and yes, I can dance, but I have never in—my *life* as a human being or as a ghost danced with a man, so maybe someday we can dance. However, I will always wish that I could have danced as a human being with one I loved, but before you ask, no, I have not found that person. Why are you interested in things so differently than your lover, Jill?"

"Girlfriend, not lover. You need a little work on your words there."

"Please explain, is a girlfriend not one's lover?"

Sam blushed and said, "No, not exactly. A girlfriend is … well, maybe both, but does not have to be your lover. One's lover must be your girlfriend or well … as a lover, well a *lover* can be, you know, *just a lover* but … let's talk about why I am interested in you and the others."

They sat and talked for a couple of hours as the evening began getting dim. They bounced from one subject to another but never touched the area of history, personal or just old history. They talked about various phenomenal ideas with the future of technology and the future of the world. Finally, Sam said, "You see, I am fascinated with things like electricity and the physics of the world. After all, that is what you are. You embody spice as I embody the mater that makes up a living being. I think there may be a way to help you in what you want."

"What do I want, Sam?" Haifa asked, looking at him.

He sat quiet for a moment and then said, "You … you want to … die."

"I want to be like you. I want to be the person as we were naturally made. Only our body dies, that is, it just …

"Rots," Sam quietly said.

"No, it just changes into simpler matter and is made into something else. It just returns to the earth. However, our soul, it just goes on. Forever and ever, to the Otherworld, or as some look at it, 'life after death.' But Sam! That is a stupid thought. Death really does not happen. You just change. Your soul Energy (sE) evolves into the Otherworld. But we, the lost ones, did not make it. There is something you must talk to Youssef ..."

Their chat was cut short with a call out from the valley floor.

"Well, we see you two have moved to higher ground with your meeting," Jill said as the three of them arrived back to camp. "Has Sam bored you enough? We had a great time looking at several of the carved tombs, storage niches, and paintings. They were once so beautiful. The faded paintings on the ceiling of the Al-Barid were so cool. They were the grapevines and flowers with cupids chasing each other in the sky."

Sam hollered back, "Well, maybe I bored her, but she will talk about the cosmos more than one of my greatest heroes, Carl Sagan."

Haifa laughed a bit and said, "Yes, we talked a lot; just don't get him talking about life when he is listening to the Doors music."

As Jill and the others got close to them, she said, "Oh, so you have discovered music and philosophy."

"Yes, we covered many things. Nevertheless, we were getting worried about you out there, as it started getting dark. Now that you are here, I guess we can make some tea. It's about teatime," Sam said as he jumped down from his rock bench.

"Well, I think we better pass on that one night," Jill, replied.

Then Saud added, "No, do not worry about drinking in front of us. We no longer think about it or have the desire. That is one of those things that you lose with time. But I do think you should start moving your camp over to the house we will be staying in tonight."

Sam and Jill looked at each other. Sam gave Jill a slight no with his head, leading her to say, "Oh, we will be fine right here. Maybe that will be good for us all. I am going to help Sam out here in our little desert kitchen. I am starving and have got to eat something. I thank you two for taking me to the painted house. It was a neat place, but I would really like to walk over every little hill and valley here. I would like to find all that the other archeologists have not found. What do you think? If I return to Petra, which I believe I will, are you going to give me another guided tour?"

Saud replied, "Yes we will guide you if we see you again. For now, we are going over to the other side of the valley and will wait for you there. There you can sleep and not be bothered by the other curious ghosts."

The three ghosts went back over to where their bags were in the small-carved building. Sam and Jill watched them go inside.

Sam looked at Jill as he handed her a cup of tea, saying, "I think we will be fine right here sitting up all night or sleeping, but with only one eye closed, and we will stay close to the water. I don't think the others will bother us. Also, I would rather the three of them stay right over there, where they are at right now."

"Do you think they want something from us or will try to hurt or … kill us?" Jill asked.

"Well, I am not sure about any of those things, except that they do want something. I know one thing, which is: we are out of here in the morning. They can follow us or stay. No, I think they should go with us."

Jill studied Sam a bit and then asked, "Why, what did you learn from Haifa? Did she say what they wanted? I sure wish Youssef would show up. I would feel a lot better then. However, he is most likely busy answering to the park guards, the police, and who knows, probably the Jordanian army. So, what did Haifa tell you?"

Sam stirred ramen noodles and said, "We did not talk about what they wanted; I just said to her that I believe they want us to help them return to their lost person, or to die, or whatever you want to call it. They want out of this world and to simply go to the Otherworld, you know, afterlife. Which I think there might be a way. I just need to get back home and to school and talk with Professor Steinhart."

Jill laid out some bread and cheese and said, "Sam, I love you and think you're great, but God, Sam. This is over your head, my head, and that nutty professor's head if you ask me. He is playing in a game that there is no winning. Let's just get out of here and regroup on how we can come back and learn more from them in a safe mode. I don't like the way you are thinking now, Sam. Are you going to tell people about our experience here? Are you going to let the ghost out of the bag?" Jill said with a laugh and then a straight serious look.

Sam studied the house across the valley floor for a bit, as he slurped his noodles and said, "I believe in Professor Steinhart's ideas and the work he has done. He is one of the world's leading scholars in artificial intelligence and artificial life, if not the best. Jill, I am glad and thankful I received my internship working for him. No, I am not going to tell anyone about what we have found. That is, until I get it all figured out. Then I will let the ghost out of the bag."

Jill looked seriously at Sam and said, "Sorry, I don't mean to sound bad toward your dreams and work. I support you all the way. Let's just promise each other that we will not talk about this to anyone without first both agreeing to who and what we say. Okay?" She then added with a smile, "Are you going to let Haifa out?" Sam flicked some noodles at her and gave a fake laugh.

They ate supper, had some tea, and set about moving things closer to the small oasis water pool. They made a bed near it with their backpacks as a couch to lean on so they could sit up and keep a watch on the valley. Sam filled all containers with water and set them around their sleeping bags. Late in the evening, as they lounged against the packs, a voice came from the slight darkness.

"Hello there."

It was Haifa's voice. Sam and Jill had their headlamps ready. When they turned them on, they spotted Haifa just about twenty feet away. Sam turned his head left and then right looking for others but saw none.

He said, "Wow, you scared us a bit. We were both about to fall asleep. Are you alone?"

"Yes, I came to ask when you were coming over to sleep with us. As I see now, I guess you have chosen to stay here. I am sorry to bother you. I guess I will see you in the morning."

Sam got up and walked out across the spring toward Haifa. "No need to be sorry, we were not totally asleep. Yes, we will see you in the morning. Are the other ghosts out?" Sam asked as he got to Haifa.

She looked at Sam and then over at Jill and said, "No, no other ghosts are out around here. They will not bother you—I think."

She turned and started to walk back to where she came from when Sam said to her, "Oh I left my BlackBerry over on the rock bench where you are staying; I thought you might want to hear the rest of what I like. So, do you sleep?"

She returned the smile and said, "Thank you, yes, we must also sleep to keep up our energy as you do. Thanks for leaving the BlackBerry out. I wanted to see and hear the others … Sam, don't go to sleep tonight, please." She then turned quickly and went to her place before Sam could ask why.

He stood still for a moment, looking out at Haifa as she disappeared into the solid darkness. As he walked back across the little spring, he turned several times trying to look into the valley. He went back to the backpack couch and got settled back into bed and stared out into the obscured valley.

"Why do they want us to sleep over there? We are fine here. They could have us bring their bags over here to sleep if they want us all together. But oh, no, they are determined to sleep over there and just keep on informing us that that's what we are to do," Jill said as Sam got tucked back in.

"I am not totally sure of what they intend to do or what they were supposed to do but see if you can get some sleep. Tomorrow we will walk out of here again if Youssef does not show up by about ten o'clock. I will stay awake to watch for visitors. If I get really sleepy, I will wake you to talk with or for you to watch while I sleep a bit if I get tired." Jill snuggled up against Sam with her arm draped across his waist, her head resting between his chest and upper arm. Soon she was sound asleep. It was incredibly quiet for some time. Then about midnight, Sam could hear soft noises out in front of him. Then the noises moved to the far left somewhere around the spring pond. It was a little wide spot in the stream about twenty feet away and about fifteen feet in diameter. Sam had placed a stack of rocks and sticks within an arm's reach of where he slept. He had tested earlier in the day throwing stick and stones with his eyes closed from his bed and could easily hit various spots in the stream or the pond. The soft noise moved to the hill just behind them. There was no water between them and the ghost, if that is what the noise was behind them. If the noise got close behind them, Sam's plan was to grab an open bottle of water and throw it over his shoulder as if one would throw salt over for good luck. He did not want to turn the light on, thinking that would give the impression that they were scared. As the noise seemed to be getting closer behind them, he also began hearing the soft noise out to their front also. It sounded like a possum or raccoon sneaking into one's camp at night, and you hear the movement of leaves. However, even if there were leaves here on the ground, the ghosts would not be making them move as they walked, unless they wanted them to move.

Sam picked up a rock in his right hand while holding Jill's lamp in his left hand, with limited movement because of Jill's sleeping body. The sound was not all around them at the same time. It was a sound that moved. It was just one thing or … one ghost. He pointed the light in the latest direction of sound and turned it on with his right hand cocked back. Just at the water's edge, not ten feet away, a pulsating figure appeared in the light's beam. Sam's right arm went forward and then back as in a quarterback's pump fake. It was Haifa.

Sam called out softly, "What, what do you want?" She stood still, but in the edges of the lights beam, there was other movement. Sam moved the beam to the left and right quickly, and then he quickly tried getting up. Jill was rocked out of her sleep as Sam got to his knees. He also turned on his own lamp, which was on his head. This lit up the area enough to show they were surrounded by ghosts. Jill jumped with a loud scream and grabbed Sam around the shoulders. He had to shrug his shoulders to give himself room to throw. This time there was no pump fake. He launched the stone in his hand at the pond, making a big splash. The dark night glittered like the aurora borealis as ghosts scattered from the spray of water.

Sam shouted, "What is it you want? Don't all of you just run off—tell us what you want?"

He got no answer. Soon, not even a glimmer could be seen in the light troughs cut into the night's darkness with the two lights. Sam grabbed a couple rocks and rushed out toward the spring, leaving Jill brandishing a club. He turned around in a circle looking up on the hill behind them and then walked over to the carved house where the spring waters came from. He could not find a single ghost.

He went over, grabbed Jill's shaking hand, and said, "Let's walk over to the where the three of them are supposed to be sleeping and ask some questions"

Jill pulled back on his hand. "No, Sam. Let's just stay here. They all ran away. That means they must be afraid of us also. Please, let's just wait until the morning."

"Fine, but you have got to quit hanging on me and grab a rock to throw in the water or a wide-mouth bottle of water to throw on them if they get close again. I just can't move with you grabbing me."

"Sorry, I am sorry, I, I just am scared. Sam, I want out of here."

"Okay, tomorrow we are out of here. Damn, they were within ten feet, and you could not see them without a light. They blend with the darkness like water blends with the ocean. If they are not energized, you cannot see them at all. So, they want us to see them for some reason."

They sat up and talked the remainder of the night, not nodding once. Just after dawn, Sam made tea and then looked all around the camp for clues, evidence, or anything that would show that the ghosts were there. He found nothing. He told Jill of the noise he had heard and was trying to find out what made the noise.

He asked Jill, "Just how could they make noise when they could walk through a bush without moving it or sit in the sand without disturbing a grain? The only way you can see them is when they activate or when they are a thimble full of spice. But I heard something."

"I don't know, Sam, but let's pack our stuff and get the hell out of here. I am not going to stay any longer than it takes to pack."

Sam walked back to Jill and began rolling up the sleeping bags, and then he said, "Let's go over and see them right now. I wonder if they are in their bags or what. If they are, we can just grab the bags and take them with us."

As Sam got up, Jill said, "Wait, Sam. Can they get out of the bags if we tie the drawstrings? What if they are not there? Let's pack first, and then we can take our stuff with us and just grab them and go."

"Good idea." Sam stopped and went back to work packing the bags, and then he said, "I would not doubt think they could get out of their bags. Maybe we should just let them activate and convince them we need to go to Amman. Bagging them may just make them mad. If we do that, we will need lots of water to … soak them. They do not like water. I don't think we can carry enough to wet them all down if they follow us with revenge on their minds. We will head straight for that little farm at the end of this canyon. There we will try to get a ride. If we have to walk all the way to Quweira, we will be short on water, so maybe we can find some there."

After pausing for a bit to think, they quickly continued to pack. They soon had all packed up and headed for the structure where the bags were put. The ghosts were in their bags. Sam grabbed one and Jill the other, and they headed out of the canyon. They did not see any other ghosts as they quickly walked down the road.

After about an hour's walk, they came to the small sheep-and-goat ranch. They headed straight for the house where there were two trucks parked. A man came to the door after they knocked a couple times. He did not speak English. He pulled the door closed as he stepped outside and motioned them over to a small work shed and pointed to some store-bought water bottles. He then pointed to a couple of benches as if he knew they were to wait for someone. Jill dropped her pack near a bench and grabbed a bottle of water. She dropped the ghost bag she was carrying near her backpack. Sam did the same but did not sit down.

He went over to the small truck, patted the hood, and said, "Quweira," He pointed at the man and then at himself and Jill. He then turned a bit and pointed to the road leading to Quweira. The Bedouin said something in Arabic, gave a slight downward nod of his head, and then walked over toward Jill.

He looked down at the bags and said, "Youssef." He walked over to the bag, slowly reached down, and picked one out of the two bags that had two inside of it. As he slowly untied the drawstring, he said, "Saud." After opening the bag, he set it back down, picked up the other two, and did the same. Jill and Sam looked at each other and watched quietly.

Jill said, "He knows about them."

Sam shook his head up and down as he walked slowly over to Jill and took a seat. They deliberated about whether they should just grab their bags and head down the road or stay and somehow convince this man they needed him to give them a ride to Quweira.

The Bedouin then pointed to the three bags and said, "English." He motioned with his hand, a slow patting 'wait' movement about two times, and then walked over to a workbench in his open shed and began repair on a tent or piece of canvas.

Sam got up, walked over to the man, tapped him on the shoulder, and pointed at the truck again. "Quweira, please, now. We pay money," Sam said as he showed the man some money.

The man looked at Sam and then over at the bags. He pointed at the bags and brokenly said, "English, me no English," and then returned to work.

Sam went back to his bench, saying, "Oh, I get it. He does not speak English, which is obvious, but he knows they do. Okay we will wait."

After about ten minutes, the three ghosts appear.

Within seconds, Saud said, "I see you have decided to go to town without waiting for Youssef. What is your hurry?" Latif and Saud stayed by the door as Haifa walked to the Bedouin and talked with him in Arabic.

She then moved close to Sam, stood still, and very quietly asked, "Did you hear me?"

"When?" Sam asked.

"In the night!" Haifa whispered back.

"Yes, how did you do it?"

Saud interrupted, saying something in Arabic to Haifa, who then went directly to the house but said to Sam and Jill as she walked by, "This is Imad's house; he is Youssef's youngest son." She went to the house as Saud and Imad continued to talk.

Jill asked Latif what was happening. She also told him that they wanted to ask the man here if they could catch a ride to Quweira.

Latif held his finger to his lips and then told her that he was going to tell Imad that he would translate to Sam and Jill what was happening. After he spoke with Imad and Saud, Latif listened a bit to what Imad had to say. He then said, "Imad is going to take you to Amman. He said that he does not agree with his father, as always. He has sent Haifa in the house to prepare some food for you, because it will take all day driving to get you to Amman."

Saud began waving his hands and arms as he spoke loudly with Imad.

As she watched, Jill asked Latif if everything is all right.

Latif said, "Yes, you will be all right. Saud just does not agree exactly with Imad and thinks he should wait for his father to come here before taking the two troubled tourists to Amman. You will be riding in that big truck over there with some goats he must take to Amman for his uncle. That is Youssef's wife's brother."

Sam looked around and then threw his arms out and said, "Just like that, Imad had a plan as if he knew we were coming. Was he waiting for us to come and ask him for a ride? If he had a plan to take us, what is he waiting for? We are ready."

After Latif explained Sam's frustration to Imad, Imad looked straight at Sam and Jill back and forth, as he talked, and Latif translated.

"I did not know you were coming this morning. But I did have a plan to take you wherever you needed to go. As you see, I do not live near my father nor do I always agree with him. I seen you the other day and talked to my father about you. I thought you might come through here again. Now I was just waiting for the three spirits here to appear so that I could better explain to you that I would take you. Okay, now, if you would please get your things and get in the back of the truck. Take this tarp to cover yourself with. You will be riding with six goats. Stay covered as we go through each town."

Jill spoke out, "In the back of the truck? Is there not room for us in the front? That is a pretty big truck."

Imad looked at her and said, "You are in Jordan, where do you think you belong? This is normal for you here. But today, I will make an exception." He looked at Latif, said something, and waved his hand to the house.

Latif said, "I know what I said to you may seem bad, but it is normal for women to ride in the back of trucks here, but today you will ride in the front. I am going to go get you something to wear. For now, just get ready to go." Latif went to the house.

Sam and Jill gathered their things and headed for the truck. After Sam helped Jill with her backpack, he reached down, picked up the ghost bags, and asked, "Do we take these bags?"

Imad said something to Saud, who said to Sam, "Yes, put them in the front of the truck, we will be going with you."

Sam handed Jill two of little bags, and they went to the truck. Soon the others came to the truck. Imad lowered a ramp for Sam to walk up into the back of the truck. He covered the backpacks with the tarp and then carved himself out a spot in the hay where he could hide. Imad and his two sons chased six goats up the ramp and closed the tailgate. His wife came out of the house followed by Latif and Haifa. She had a basket of food and a bundle of woman clothes for Jill to wear and cover her head and face. Jill put the black robe on over her cloths and got help from Imad's wife putting the headdress on.

Chapter 4
Goat Truck

It was not noon yet, and they were finally headed off to Amman. Latif and Saud returned to spice in their bag. Haifa rode in the back with Sam and the goats. Sam sat on his cushion of hay and stayed covered with the tarp through the towns, especially the first one, which was Quweira.

Jill could not talk with Imad up front, so the first ten miles or so were spent in silence. When they arrived in Quweira, Imad drove close enough to his father's house to see if his truck was there. Not only was his father's truck there, but there were two ranger trucks as well as one police car. They did not even slow down. Imad sped up a bit and watched the rearview mirrors for several minutes even when the house was well out of sight. He slapped the dash, getting Jill's attention, and then rattled a string of words she did not understand. She shook her head, agreeing with him as he continued to jabber and point back toward the house. Knowing she could not go all to way to Amman without talking to Imad, she loosened the bag ties, hoping Latif and Saud would appear to translate for her.

In the back of the truck, Sam threw the tarp aside, also hoping Haifa would appear. Within a few miles, they were both honored with the presence of their new ghost friends.

Jill soon found herself with Latif sitting between her and Imad. Just as he appeared, Jill asked, "Will you please explain why Imad is not happy with his father? Also, why is Imad so intent on taking us to Amman?"

Latif told Jill to be patient for a few moments while he explained to Imad what he was going to do and ask him if there was anything he wanted to say.

After the two talked with each other for a few minutes, Latif said, "Jill, I am sorry for what I must tell you, but you must understand, we have lived an exceedingly difficult life as misplaced souls. You must understand that we left our dying human body as all souls do at that time of separation but could not exit our entrapment to the Otherworld as most living souls do. Why, we do not understand—yet we watched our human bodies decay. It is so very punishing to see yourself crumble and deteriorate when others have not, and you have done nothing to warrant this. It was so sad and frustrating at first, but we soon learned that the body is just a borrowed structure made of matter like all else around us. What we could not understand was why we could not find our way to where we were to go after we leave this borrowed structure. We have the power and the knowledge for many things yet could not find our way to the Otherworld. So here, we are for eternity. Our biggest mistake was to be led astray by our religion, as we knew it. When we realized we were wrong, we followed the next teachings and the next, on and on until today, and as you see, we are still here. But also, now I am deeply sorry for what we have done."

Jill stopped him and said, "Latif, you owe us no apology, this all was as much fault of ours as it was yours."

"Yes, it was our fault. We put you in danger."

"What are you saying, Latif?" Jill asked as she leaned over a bit to look at his face.

"Okay, you see, Youssef was a trader of spice in Petra," he said as he stared toward Jill's feet.

"You mean he *is* a spice trader."

"Nooo, today he is a herder, but in Petra he was a very prominent spice trader when he was a Nabataean, not just a Bedouin herdsman like today."

"I don't get it. How can that be … Wait! Don't tell me that … he is …"

Latif slowly nodded his head and then stared out the front window as Jill continued to try to look at his face.

"Yes, that's right, Youssef was a ghost, but now is not. You see, he, Saud, Haifa, and I were all entombed together. Together we flowed out through the porous walls and crumbled into the sand over time. We are what cause the colorful walls to slowly flake off in the structures of Petra. The spice pile we died in gathered our soul's power and that is why we carry the scent of a spice. Then, just a few years ago, something strange happened to Youssef, and he became a human, a person again. We did not know how this happened. There were some of our people who were able to escape the tombs before us who told us of some ideas they thought happened, but we were not certain of what happened."

Jill asked. "What do you mean; you were not certain what happened? Is that what happened to Youssef, or your past life in Petra?"

"Both. There are stories or tales that tell how the first appearing ghosts washed out through some cracks in the sealed tombs into the sand. They were angry of course and took revenge on the Romans who had moved into their homes. They were able to snuff them out and …"

"You mean kill them, suffocate them?"

"Yes, they figured out a way to gather themselves over the Romans as they slept and suffocate them. It is said that then the Roman's soul left his body and went to the otherworld, the Nabataean lost soul overtook the body and lived again until he died naturally or however, and then he got to go to the Otherworld.

"So, they both today are together in the *Otherworld?*"

"Yes, I guess, but we know nothing about that because—well, because the Otherworld is there, and we are here. Maybe they are still fighting there, maybe they are friends. We do not know."

"Youssef told us what happened to make him a person, and he wanted to help me, and Haifa become a person again like him."

"So how … did he become a person again, or do I want to know?"

"You are not going to want to know, but this is something you must know. One day, a Bedouin herder found this pile of spice near the entrance of a structure. It was Youssef. The herder gathered up the spice in a bag and took it with him. He knew what it was and knew that he could use the ghost in ways to help him. That night, the herder went to sleep out in the open near his herd. He used the bag in which he held Youssef as a pillow. During the night, a rain came, and Youssef tried to escape the rain by activating and running away. You see, we do not like water because it deactivates us if we are in the ghost form or just keeps us a like a mud until we become dry again. So, in trying to escape, as he became a ghost, rain deactivated him. He fell into a web of spice over the Bedouin's face and smothered him. The Bedouin died, of course, and as he did, his soul exited his body to the Otherworld. However, this allowed Youssef's soul to occupy the herder's body."

Jill put her hand to her mouth and gasped, "Oh my God! A transplant. He simply transplanted into a body becoming a person again. Reborn."

"Yes, that is about what happened, no harm done. You see, in the end, they both are allowed to go to the Otherworld. We have learned over the years how stupid and mythical all religions are. Like yours is as untrue as ours was. When someone dies, all become sad and have such a gloomy and depressed funeral when they should be celebrating. People have even written and forced their children into believing these myths and teach how great it will be to go to heaven but are all gloomy about going. Is this not stupid?"

"Yes, you are very right about our mythologies, but I am still not quite sure about the 'no harm done' thing. You see, I am not ready just to pop into the Otherworld yet. But—I am not afraid to take that change, and I have someone waiting there for me."

Latif sat still, looking down at his hands folded together in his lap, and said, "Yes, I also have my mother waiting in the Otherworld for me, but I am not sure I will ever go there."

Jill said, "So today Youssef is a sixty-year-old Bedouin who once lived two thousand years ago, who became a ghost when he was younger. Now he will die and go to the Otherworld. So, you do not age as a ghost?"

Latif answered, "Yes, we continue through time at the same age forever. There is no end of time or the soul, which sounds good and is good, that is, if you and your loved ones and soul mates stay together."

"So, what do you mean by that, if it stays together?" Jill asks.

"Well, I think you should talk with my sister about that. She is much smarter than I at that, and she loves to talk about that kind of stuff."

Jill sat silently for a bit, and then she looked through the back window of the truck and said, "Your sister—you mean Haifa?"

"Yes, she is my twin sister."

"Wow, that's interesting, so what happened to your mother and father? Do you remember them?"

Latif continued looking at his hands as they molded around each other and then said, "Yes, I remember them, but first allow me to tell you what the plan was for you and Sam. You see, Youssef always knew what had happened to bring him back to a person, but always kept it a secret. He said he felt bad for many years and stayed alone. The other ghosts and I did not know what happened to Youssef the ghost until he came back to us. So sometimes, a ghost just disappeared, and it is never seen again, like Youssef. His soul simply took over a human's body. The human soul was released to the Otherworld as naturally happens to all animals. That is what happened in the old times with the Romans—they began to wonder off from Petra into the desert. The Romans body was taken over by the Nabataean's soul. The Romans soul went to the Otherworld and the Nabataean went to the desert to start a new life. Today it is taught that the Romans were deserters. However, what really happened was that the ghosts first released by the earthquake flooding the tombs simply snuffed a Roman who then went to live in the desert as a herdsman or whatever. They learned that they could suffocate the Roman soldiers and leaders, thus allowing their soul to exit their body structure. Some reborn ghosts did not go to the desert. Instead, they became very revengeful and killed as many Romans as possible by capturing them and snuffing them with a spice ghost until they were out of Romans. That was not good, because it was a waste of body structures as other slow-coming ghosts needed to transplant into."

Jill asked, "So why did it all stop? Why are you still here?"

"Like I said, we ran out of Romans. We also did not learn the way they had done it when we, the slow ones, seeped through the stone walls as you see colored today from the color of spice."

"Can this still be done today … or is this still being done today," Jill asked as she tilted her head down in order to look at Latif's face.

"Yes—it can be done but has not been done for hundreds of years except with Youssef. We cannot do it by ourselves, because we have to be watered down and in the right position. We cannot pick up a bottle of water and pour it on another or ourselves while a live human being is sleeping. We would need the help of one who is a living human being."

Jill then asked, "And so that is where Youssef comes in?"

"Yes, he was going to help me and my sister Haifa to become human beings once again so that we would be able to continue on to the Otherworld as he was soon going to be going … Jill! Youssef is—my father."

Jill looked straight out the front window in silence as Latif continued looking at his lap. His half-brother, Imad, looked over at Jill with a blank face, nodded his head slowly as in 'well, now you have it,' and then looked back at the road.

Jill turned and stared out the side window as the truck lumbered down the road.

In the back of the truck, Sam was been joined by Haifa. She emerged from her bag, settled down in a cross-legged posture like Sam, and said, "Hello, well, you are on your way to Amman, where we will say our goodbyes. I hope you will return here some day in the future. I have enjoyed talking with you. I enjoyed most all of your music. You did retrieve your device."

"My BlackBerry, yes, I have it here with me, fully charged."

"I have listened to thousands of tourists and local people, but do you know, Sam, that you are the first human being or non-ghost that I have talked with since I became a ghost?"

"Why is that? Why do you not talk with others or show yourselves to the world?" Sam asked.

Haifa appeared to lean back against the cab of the truck and look out over the tailgate as they drove down the road. She said, "Because that just would not go over very well. Come on Sam, we are a spirit—you know, the soul—and we really do simple exist. However, to people, that is live human beings … oh, we are a paranormal thing, spooky and evil. Now, do you think we could just show up at the downtown café and have a good old chat with human beings?"

"Well, I guess not, unless it was Halloween," Sam said as he laughed a bit.

"Also, I think it is because we would be looked at as the one who it seems to live on forever or are what is thought of as immortal, where we really are one and the same. It has come to me and the others here that life goes on forever. Life goes on forever in the soul, but as human beings, we thought of it as separate two things, our life, and our soul, where it's really not. After two thousand years, you would think the human-being part of the soul would get smart enough to figure that out. But they cannot because they are too pacified with their fabricated religions to answer all unknown questions."

Sam laughed aloud and said, "Wow, what a philosopher I have met, a ghost at that. So, are all of you here philosophers or as smart as you?"

Haifa also laughed a bit and said, "No Sam, they are not all as smart as me. *I am the great wizard.* No, seriously Sam. We are all just a little different from each other, just as you as human beings are. Some continue to study the world and seek out answers all the time, while others stay stuck in old stupid traditions and even kill others who do not follow the majority tradition. I am a seeker of knowledge and have learned by hiding around in the hills and listening to tourist talk. We have the chance to learn many different languages and people's thoughts throughout history. We have seen just about every different race and culture on this here earth. Which really are both a waste of time and thought for the human being."

Sam interrupted asking. "Now why do you say that? Those are two particularly important elements within human civilization."

"Because, Sam, over time we here have learned that really, our human-made cultural inventions are the greatest cause of war and destruction. We, as ghosts, have come to realize that each and every human should be classified as individuals not groups. DNA codes, after all, are an incredibly unique individual blueprint of its own... Now we cannot read our DNA code as you humans can. However, we use the idea that our race is that of an individual, not a group. As for culture, it is one's own talent, your job, your work. On the other hand, something like your hobbies—not just your language or your religion. Individual talent rather than social talent would not lead to a social disease as we see so much of today only to be combatted with war. I think in about one hundred years, human beings will begin using DNA as their race, not their body color. Colors are like numbers from zero to the infinity, there are no two the same, and there is no end or limit but in one's own vision. That is, unless you just reproduced an exact copy of one. As for making a biological being, this could be done, but no matter what is made the law of change is always taking place. Two clones made on the same day will not be the same one hour later."

Sam agreed, shaking his head slowly, and then said, "You are right, and I bet have seen and heard of many horror stories over the years. However, there are a few human beings advancing out of old traditions, slowly. Each generation there are more open-minded humans coming out of the closet."

"So, do you think there will be a time when humans will create life? Or let's say, transplant life? After all, humans are transplanting parts of life, such as a heart transplant. But of course, that is not what is thought of as pure life."

Sam's eyes lit up because this thought hit home for him. "Yes, I think that is going to happen very soon. Actually, I began working for a professor last fall as I began my graduate studies at MIT in that exact field. He and several others around the world are working on this great endeavor. Who knows when it will happen, but I think it is just a couple years if not months when it will be done? As some say at school, we have pots brewing right now."

"Sam, you know, in the short time we have known each other, I have to tell you … that I really enjoy being with you. But there is something I want to tell you and Jill. That is, I am sorry for all of the trouble we have caused you, and I hope you will forgive us for this trouble. I wish I could communicate with you as humans do, and as I once did. You know, such as letter writing, telephone, and things like that. I miss not fulfilling my human life. I am sure that would have been better than getting stuck here in limbo."

Sam looked at her and said, "Ah, you did not cause us much trouble, and now we are on our way to the airport. Sometimes you had us a little scared and worried, but all is well now. Heck, we are just a couple of young people trying to find our way in life out gallivanting around, living out of a backpack, and really loving it. But I must say now, it has been the most exciting time of my life. I will come back and look you up. When I come back, I will go to Little Petra and camp there, wait for a ghost to appear, and tell them I am looking for you. Will that work?"

"Yes, that will work. I will pass the word to all to watch for you. But I think they all are aware of you now, especially the guy you almost took out with the stick."

"Oh yeah, tell him I am sorry. Is he all right?"

"Yes, he is fine, the water splash deactivated some of him and that part of him fell off into the sand, and it was retrieved later. That happens all the time when it rains."

"So how did you make the noise giving us a warning?"

"I went around before the others and scared up some desert hamsters, causing them to make a lot of noise as they scurried around. Okay, now a bit of the plan for today. I think you will be going to the house of Youssef's brother-in-law. You will wait there while Imad goes to the airport to buy your tickets."

The truck turned onto a sandy road as Jill stared out the passenger-side window. She had not said a word for an hour. She turned to Imad and asked, "Why are we turning on this dirt road?" Imad looked at her and then at Latif. Latif translated, and told Jill that Imad was taking a short cut over to another paved road.

Latif explained, "He said that we will arrive in Amman in a couple more hours. We will stop at his uncle's house, where he will leave you while he goes to the airport to get your tickets. At the house, he says you will need to give him your papers and money. He will buy tickets to Tel Aviv, leaving as soon as possible."

Jill told Latif to tell Imad thank you and then asked, "So who and where is your mother? I guess she also became a ghost?"

"No, my mother died as a human, as a captured slave. She along with her mother and father were put under house arrest by the Romans."

"So, what did she and your father, Youssef, do when they were Nabataean people? Were they spice traders or town officials?"

Latif lifted his head, looked at Jill, and said, "My father and mother were especially important people back then. I was raised by my mother in the most beautiful home in Petra. My sister and I were given special classes in school as well as in sports like horse riding. We were taught the skills of arts and crafts. Our mother looked like Haifa; she was thin with long dark wavy hair. Her face was sleek and the color of cinnamon. She had eyes shaped like almonds, light green in color. Full lips surrounded her perfectly shaped teeth. Her name was Katrina."

Jill looked toward the back of the truck through the rear window at Haifa. Then she said, "She must have been beautiful, but it is difficult to look at you and your sister and see what your real features were like. I am sorry for what you are going through, but that does not give you the right to take anyone's life just to fix yours."

"Yes, you are right. That is why together we chose not to do, as our father wanted us to do. We went to Imad for help. My sister gave you warnings during the night. Now you will soon be on your way back to your mother and father and your great life."

Jill finally looked back at Latif, saying, "I also went to a special school of music. That is where I was when she was coming to get me one day. But that day she never picked me up. My father picked me up, who then informed me that my mother and the love of his life had just been killed. My mother was killed at the hands of a drunk driver. Therefore, I have also lost my mother. I understand what you have gone through. That still does not give your father the right to do what he was doing."

"Would your father not do the same? Or you, for your daughter or son?" A long silence took place while the truck bumped down the sandy road.

"I don't know," Jill said.

The truck rambled on down the road for a while until Imad came to an intersection where they could get back on the paved road. They stopped there for fuel and a break at a lonely service station. Haifa and Latif went dormant to stay out of sight while Jill and Sam used the bathrooms and returned to the truck. In their short conversation, Jill told Sam that they needed to talk but were cut short when Imad returned from his break. They loaded back up and continued to Amman.

Latif reappeared and asked if Jill needed any translation from Imad.

Jill said, "No, I do not need anything from him. So, when did you learn that Youssef was your father? How did you like it knowing he is a reborn human?"

"We knew that he was gone for a while, and that is normal. You see, we can lay dormant in our bags for days, weeks, sometimes for months. We do not think of time as you do or as we did as humans. When he returned as the Bedouin and came around our most popular gathering spots, we quickly learned who he was. The body he was in we did not know, but the person we knew was Youssef, our father. He told us all what he thought had happened. He then had to resume the life of the Bedouin whose body he was using. He was lucky—the man did not have much family, but he did have a small ranch and a herd to manage. Our father was smart and a good herder, so he became an extraordinarily successful Bedouin. He then married a woman who had been arranged for him and had other children, as you have seen. We have always gotten along together, but it seems as they get older our half-brothers and sisters grow more distant from us."

Jill said. 'What do you expect, Latif? They will die someday, and you are immortal. You live on forever, I suppose. This is the worst event in all human beings' lives. We fear death. Well, I guess I should say it is almost the worst thing we fear—for me, the worst thing was losing my mom. No manmade church can replace my mom, and I will never accept wishy-washy answers claiming they have it all figured out. Without the word *faith*, preachers would be in the unemployment line. That is the only thing that gives preachers a job—a high paying one for some and poorly for others preaching about the same God—that is, to live off the fear people most dread as well as other misfortunes we have in life. Then there are some who don't even need the preaching. They just invent another small group brainwashing machine. Okay, enough preaching. So, your father Youssef is Muslim, right?"

"No. not hardly, he just tells people that because here in this country, it is not wise not to be."

Jill finally laughed a bit and said, "Nobody would believe this conversation. I do not know what I will do with what I have learned here. If I were ever to talk about it, the know-it-all church people would either pray for me or burn me at the stake. The scholars would put me in the hospital or a psych clinic, but I am a person of the spirit and now a spirit has enlightened me. I do not follow any religion anymore. I agree they are all manmade; however, that does not make me an atheist, so let's get on with life. After this trip, I have become wonderfully comfortable in what I have learned yet am really in wander of what will happen next. After this trip, my life will never be the same. What will you do now that you are not going to change or transplant bodies with Sam and me? What a question, right? You will not join your father in the otherworld, and yes, he soon will be going there. I will see him there someday, and you will not."

Latif had nothing to say for a while, and then he said, "When you get old … I ask you to come here and take a message to my father and mother for me. However, the message will really be the same as today—that is, I would say hi, and I am doing fine and the same as always. It will do me no good to ask how they are doing or what they are doing, for I would get no answer. I wonder what it will be like when my father finds my mother, Katrina, there."

Jill then asked, "So do you know where your mother was buried? Do you go to her grave like we human beings do?"

"No, I do not know where she was buried. Yes, I am sure I would go to see her human grave as human beings do. But this would only serve as a reminder of her to me, but I will never forget her even without reminders."

"So today you have nothing of her, not even one keepsake."

"No, nothing, except the ruins of our house. Here today it is called the castle of the pharaoh or king's daughter. She tried to give me and my sister each a bracelet she wore as the Romans stripped her away from us. But it all happened too fast. We, along with our father, fought and cursed the Romans, finding ourselves sealed in our tomb that same day."

"Does your father know where she was buried?"

"I do not know. He does not talk about her. But he did say that there will be a time when we will all would be together."

Jill told him that if she could, she would like to return to Petra and learn more about it. She told Latif to try to find out all he could, and maybe she could someday help him find where his mother was buried.

Chapter 5
Amman

By early afternoon, the truck came into the large city of Amman. They went straight to the old downtown area where Imad's uncle lived. He also had a small sheep and goat farm outside of the town but lived-in town most of the time. His farm was where most of the family brought their livestock to sell. As they all unloaded from the truck, Jill grabbed the two spice bags from the front seat. Sam got their things from the back of the truck and the bag with Haifa. They were quickly taken into the house where they met the uncle of Imad and his wife. They were expecting them and showed Sam and Jill a room they could rest in until they departed. Imad left to take his goats to the ranch, and then he would go on to the airport to get the tickets for Sam and Jill. He gathered their passports and money and told them that he would get them a flight out that very day or as early as possible. They were to stay with his uncle's wife in the house until he and his uncle returned.

Just after they left, Jill opened up the ghost bags, hoping this was a sign to them that they needed to appear. Within minutes, the three of them appeared. Without saying hello, Jill asked how far it was to the Jordan Archeological Museum.

She said to Sam, "I know the timing is not right and that we do not have much time, but I cannot leave this place without saying that I at least stood at the base of the great pillars of the Temple of Hercules. That temple is at the top of Citadel Mountain near downtown Amman. We can catch a cab right down at the street corner, go see that temple, and then go down the hill and see the Roman amphitheater before we have to leave today or tomorrow."

Sam responded with resistance, saying, "I don't think that is a particularly good idea. I know what it means to you and that it was one of your dead-set destinations, but now the most important thing is getting out of this country. I say we should just patiently wait right here."

Jill turned to Latif and told him to ask how far it was, and how long it would take to get there by taxi. After talking to his aunt, Latif told her that it was only minutes away and would be quite easy. Within minutes, she had her small pack in hand. As she gave Sam a kiss, she said, "We still need to talk about something important when I get back."

"If it is important, don't you think now would be a good time to talk?"

She went to the door and was on her way to the corner where a taxi went by about every minute or two. Latif asked to be taken along in case Jill might need help. Jill grabbed his bag just as he returned to it, and they were off. Sam went to the corner of the living room and found himself a seat up against the corner. Haifa went to the kitchen where the uncle's wife was working. Saud joined Sam on the floor carpet.

Sam looked at Saud and said, "I wish I could take you back to the laboratory at MIT and study you along with others there. I think you are one big ball of organized electrons with a small bit of 'matter of spice' at times," Sam said with a grin and small laugh.

Saud gave a surprised stare toward Sam and then said, "I am glad to know whatever it is I am now made of. I also know I can outrun or out-move you, because I am beginning to think you would like one of us as bad as we wanted you. But I don't want to go to your lab; I will see what I can find for you, though."

He moved a few feet away as Sam said, "I think you gather your energy from the surroundings, that being the air in which we breath and or anything you come in contact with. Are you sure, you wouldn't like to go back to the States with me? You would be the biggest celebrity in the world."

Saud hovered in a cross-legged position near the wall. He looked at Sam and restated his feelings. "We are in no need to be studied. We know who we are. We are the soul of a person, a misplaced soul only from our soul mates, not from our independent conscience. I think it would be better for you to just study your own soul while you still have it as it is. One's soul does not change through time, other than becoming more experienced. We can see human beings still living and employing the same old threats of punishment to control people's behavior and ways, robbing, or limiting their freedom to live as they want. Now, if you want to become just your soul, I can help you with that. Then you can go see yourself as I am."

Sam quickly looked at Saud and also decided to move a bit, and said, "I think I will go see what they are doing in the kitchen. Besides, I smell food." He got up and went to the kitchen where he found Haifa up on a counter watching the lady preparing things for dinner, and a sack lunch for Sam and Jill to take with them. It was always a thing of the Arab people to drink tea before they talked at a gathering and always a proper think to feed visitors. Sam asked how long they thought it would take Imad to get the tickets. Haifa told him it would only take a couple of hours. She told Sam that the lady of the house wanted him to take a seat at the bar counter and have something to eat. As Sam moved to the counter, where he sat on a stool, the lady of the house brought him a glass of pomegranate juice mixed with fresh lemon and orange-blossom water. As he sat talking with Haifa, the lady brought him a shawarma sandwich made from shredded chicken with a bit of diced cucumbers, tomatoes, and onions. They were all wrapped up in fresh pita bread and spiced up with a sauce mix of garlic, allspice, tahini, and yogurt. Sam again apologized to Haifa for eating in front of her.

She thanked him for his thoughts and said she wished she could join him. Instead of bothering him, she went to join Saud in the other room.

As Sam finished eating, the lady spoke to him in Arabic and pointed to several things she had made, and then she pointed to Sam. Sam nodded his head in thankful acknowledgement. He finished eating and went to the living room. There he saw that Haifa and Saud had both returned to their bags. Sam carried his and Jill's two backpacks to the back room where Imad had said they could rest. As he took the second bag back, he motioned to the lady of the house that he was going to take a nap. He pointed to the desserts she was making and the sack lunches sitting on the counter, and then he pointed to his backpack and smiled. The lady smiled back and motioned that she would fill his backpacks with foods for many days. He looked out the window down toward the corner of the street where Jill had caught the cab. Slowly shaking his head, he walked to the floor mat, grabbed a pillow, and stretched out for some rest and an unpredictable wait.

At about the same time Sam kicked back to get some rest, Imad and his uncle reached the airport. Jill also had just arrived at the highest point overlooking Amman and was dropped off at the Citadel Mountain tourist center. This is where another incredibly old city was built in the beginning of human civilization. There were several legends of this being the home of Hercules. On top of the mountain was one of the largest temples ever built in all of the Middle East. Here is where the two tall columns stand today of the remains of the Temple of Hercules. Jill quickly went through the small museum there and ran over to the columns. She just had to wrap her arms around one of them and get another tourist to snap a picture of her. Thanks to the power of ghost, recharging the camera was now in working order. She walked over to the lookout point where one could look down at the Roman Amphitheater and its museum in the heart of old Amman. She took a couple more pictures and then walked toward the taxis waiting for a rider. She did not have to walk five feet before she was hit up by several taxi drivers willing to take her to the many different places to be seen. They did not seem to care where she wanted to go but had many tourist attractions, they would gladly take her. To get away from them, she turned and walked over to a cage-like door made of steel bars. She grabbed one of the bars shook it and said loudly.

"I want to go here, in here. Can I go in here?" This attracted the nearby armed guards that patrol every tourist site in the country. As the guards, approached the taxi drivers quickly scurried off to their cars.

One guard said, "May we help you? This door is locked, and nobody is allowed there."

Jill smiled really big at them and answered, "I did not mean any trouble. I just could not get those guys to quit telling me where they could take me. I simply wanted a taxi to take me down there to the Roman amphitheater. Is that where this door leads to because I can see a steep stairwell?"

The guards came over to the door, looked down into the deep darkness, and said, "Yes, that is the old secret passageway they used to transport animals, people, and things used in the shows down there in the great amphitheater. It is still passable today but too dangerous to allow people into. Come with me, and I will take you in my car down to the theater."

Jill followed the guard to his car as the other guards stood still or went back to their causal walking around. As they got close to the car, Jill asked, "How much will this cost?"

"It will cost you nothing. I need to go down there to the office, and this will also make my friends you see there, a bit jealous. You see them there watching. They love to help out young American ladies and would love to flirt with you as well. But I tell you—if you encounter this event, cut it short, and do not talk with them. It was a good thing what you did."

As they drove down the mountain, Jill got a ten-minute course on the history of the area. Then at the amphitheater site, she got a free tour of the large theater through its grand conglomeration within the great stands and the hidden underground passages and rooms, by the guard and a friend of his working in the museum. When the park was closed, the guard helped her get a taxi back to the address she had written on paper. She asked the guard to explain to the driver to let her out at the corner just before the house address. That way she could see if any of the police or Youssef was there. It was beginning to get dark when she approached the house. After the driver let her out, she began to have a bad gut feeling. As she got closer to the house, her feelings became real. She could see Imad's large truck, several other cars, including Youssef's truck, and a government official's car in front of the house. This was not in their plan, but just like about every day for the past few she had to come up with something. She did not stop but veered off the battered sidewalk into the side bushes and continued toward the house. No one was outside, so she eased her way around the back of the house where she could see in the windows. She could see and hear a large group of people in the living room but could not see Sam, so she crept around to the one and only window that was in the room where Imad had told them to rest. Looking through the window, she could see Sam sitting on the floor with his hands folded on his head and his head resting on his knees. She gently knocked on the windowpane to get Sam's attention. He quickly raised his head to look at the window. His face lit up as he crawled over to the window, got up, and unlocked it.

As he unlocked the outside bars, he said, "Man, am I glad to see you. They are waiting for you to get back to haul us in or something. I guess it's a good thing you went to the museum."

"Well, tell me later. Hand the packs out and let's see if we can escape one more time."

Sam handed the two backpacks out the window and then climbed out himself. They scurried off to the corner to catch a cab.

Sam asked, "Do you still have Latif?"

"Yes. Did you get our papers and tickets?"

"Yes. Just as Imad came in the house, he gave me our stuff and sent me to the room. He said for me to stay there and not come out. He told me that the authorities were not allowed to go into any other rooms according to the law. He said they would have to leave and then return with the city police to enter again. When he got rid of them, he would help me get out and hopefully be able to find you. We have a red-eye flight out, so let's get to the airport."

Just as they were about to flag a taxi down, everyone came out of the house. Jill and Sam were quickly spotted flagging a taxi down. Everyone from the house scattered to various cars to take pursuit of the taxi. Just as they got into the cab, Sam leaned over the front seat and yelled to the driver to hurry to the airport.

After looking in the back window, Jill could see the cars turning around in the road and headlights coming on, but no blue flashing lights. She grabbed Sam by the belt, pulling him back into the rear seat.

She then leaned over the seat and shouted, "To the Roman amphitheater downtown."

The cab driver turned around, asking, "Are you sure? The Roman theater, it is closed."

"Yes, to the Roman amphitheater or theater, whatever it is called, we need to go get something we left there my purse—I left my purse there with all our money in it."

The driver turned back around and sped off, leaving the chasing party behind but not totally out of sight. Jill knew exactly the price the cab would be, so she dug in her small pack for the taxi fare, getting ready to chuck it to the driver and run, knowing the drivers always squabbled over the fare. When the car came to a stop at the museum front entrance, Jill handed the money over the front seat and dropped it. They grabbed their backpacks and Jill led the way around the building. As they ran, Jill opened Latif's bag. This worked with perfect timing. The taxi driver was out of his car yelling that he needed more money as the chasers pulled up behind and in front of his car. The taxi driver quickly attacked the pursuit gang, demanding help from the man in uniform, Raul. The driver grabbed Raul's arm and yelled for help in getting the money he felt he was owed. This gave Jill, Sam, and blossoming Latif a little more time. As the three of them ran full speed, Jill gave Latif instructions to unlock the gate that went into the amphitheater. He did this without hesitation and relocked the gate after they got in. Jill told Sam and Latif to follow her as she ran through the winding small hallways of the old seating stands. These all had been covered at one time but were just an open looking maze now. As the heavy breathing two looked back, they could see a stalled confusion at the locked gate. This gave them more time to carry out Jill's sketchy plan.

Outside the gate, police cars and army guards began arriving to investigate the unauthorized visiting. The gate was reopened, and the pursuit continued. Raul came to Amman along with Youssef and was in the lead, telling the police that the intruders had escaped from being questioned at the Petra park. He made it clear to all that the two unidentified tourists were under his jurisdiction and needed to be taken back to Petra for questioning but that no major crime had been committed. When Raul saw that Jill and Sam went down into the underground chambers of the stage floor, he stopped and told the police to let him and Youssef along with two police officers chase them down. He told the other police officers to go outside and surround the place. As the group of police started back out, Youssef told them to find the light breaker box and turn on the park's lights. As the large group of police left, Raul informed the two police officers with them that these were escaped prisoners from

Petra, and that he and Youssef would be the only ones to take them and question them. The four then began their search into the dark underground maze with the aid of the two flashlights the police had.

Jill, Sam, and Latif soon came to another locked gate. Latif put his hands on the gate's lock and stood still for a few seconds. A click and a small rustling sound came from the electronic lock. The gate then gently swung away from them with a little squeak. They went through the doorway into a room filled with columns that supported the theater's stone floor above them.

Jill pulled Sam and Latif through the darkness until they heard Youssef talking to Raul not far behind them. She suddenly realized they did not relock the gate. She stopped and pulled the two with her behind a column. In the total darkness, it was easy for her to tell whose hand was whose. She pulled Latif close to her lips and whispered into his ear, "Sneak back to the gate and lock it so no others can come in." When she lowered her cupped hands from the side of Latif's head, she had to shake them to free the tingle. As she did, goose bumps ran down her spine, making her shake her head and shoulders. As she talked with Latif, Sam dug into the top backpack pockets for their headlamps. She then whispered her plan to Sam. "I know where the tunnel is to the top of the mountain. There are many other tunnels down here, but on my little tour today, the park guard and his friend showed me where the old secret tunnel was. I hope they all just stay confused and search for us down here. When Latif returns, we will head for that gate and try to sneak through without them knowing where we went. We can then catch a taxi up at the Citadel Mountain part of the park."

Sam retrieved their two headlamps just as they heard the door slightly bang shut. The two police officers had flashlights that could be seen racing toward the door that Latif had just relocked. They did not want to turn their lights on because they were too close to the door to be seen, so Jill just waited to feel the return of Latif. Soon an impulse ran through Jill stimulating her to reach for hands and run.

"Shhhh," she said as she grabbed the two hands again and pulled them into the direction of the tunnel. When they got to the far end of the room, there was again a locked door. Jill guided Latif's hand to the lock. As she looked back, it became obvious Raul and Youssef had obtained a key to the doors. Lights were coming toward them from their backs but were slowly searching from side to side. The door unlocked with a loud click and they rushed through. Jill turned and slammed the door shut with Sam's help. However, it was too late this time. Youssef grabbed the bars of the door in the darkness and shouted loudly. "Stop and we can talk; there is something you must know."

He and Raul had left the police behind searching around each column, knowing where they were most likely headed. Again, a voice came through the bars between the struggles. "Stop this nonsense, you cannot escape the authorities. Let me help you. We will all talk this through," came Youssef's voice.

Jill pushed her face up to the bars and yelled, "Talk what through? What were you going to do with us? Talk through our death sentences? Does Raul and the police know? Before we give ourselves up, would you please explain all of this to your three helpers behind you and the many other that will be coming?"

As she angrily shouted through the bars, Jill felt a tingle at her waist. She and Sam pushed hard on the door and then heard a click. The door was locked, and she felt her whole side tingle as Latif must have brushed up against her as he locked the iron door. Jill continued screaming as Sam pulled her back away from the door. Latif eased off into the tunnel knowing where they were headed.

The two police with the light quickly ran to the confrontation they could hear. When they arrived, one of them yelled for Youssef to get away from the door. Youssef turned, looking into the police light beam with his hand out, saying, "Wait, let me talk with them. I think …"

A shot was fired. A bright flash lit the room up along with an ear-blasting bang from the police officer's gun, causing Youssef to slowly slump in his shadow from the yellow beam.

Raul yelled at the police, telling him to not shoot again, and went to Youssef. Through the partially obstructed light beam, he could only see Youssef's hand as it guided him down to the floor where he then lay almost motionless. His left hand was still clinging to one of the bars, easing his decent, as the other turned red where the bullet penetrated his chest. His head and shoulder rested against the bars at the bottom of the door as he rolled over on his back.

Jill rushed back to the door, knelt down, grabbed Youssef's arm through the bars, and said, "Why? Youssef, I understand your longing, but it was not right."

Youssef looked into Jill's eyes and said, "I know I was wrong, but I wanted us to be together. I wanted my son and daughter to be able, to soon join their mother and me. This was the only way I could think of to get them there with us some day. Now it will only be I who goes to the otherworld, leaving my two loved ones behind forever. Please help my two children to find the tomb of Katrina for my children's closure. Raul will help you because he also wants to find it. He does not know where she is, nor does anyone else."

Raul and the two police officers argued about what to do, leaving Jill shaking and Youssef becoming more still.

Youssef said to Jill, "She—with mother and father the king, were cooped … caged in …" Youssef's chin fell down against his chest as his eyes slowly winced in pain.

Tears started flowing from Jill's eyes even though she still had some anger in her heart. She shook his hand, asking Youssef to talk. "In where, Youssef? What are you saying, Youssef? Please don't go now! Katrina is where?" She choked and continued to shake his softening arm through the bars. She tried to but could no longer talk. As Youssef's hand released the bar, a small piece of paper fell from his hand.

As the police and Raul came toward Youssef, Jill grabbed the bit of paper and then shouted and cursed at the police and Raul. "Are you happy now? Why did you shoot? Were you trying to shoot me for trespassing, or what? He was only trying to help you capture my friend and me. Are you going to shoot me now? You might as well you…"

One of the police rushed forward, making a grab for Jill through the bars. Jill sprang away from the gate, falling on her butt, and kicked the police officer's hand back through the bars. She scrambled out of reach and to her feet, and then she darted off into the dark tunnel to avoid another shot that might come. Sam and Latif were not far along in the tunnel. When they discovered Jill was not with them, they stopped and began a slow backtrack.

When she caught up to them, Sam said with alarm, "Jill, where were you, we heard the shot in the dark and ran faster, keeping the lights off so they wouldn't have a target. We then noticed you were not with us. Are you all right? Did they get through the door?"

Jill answered as she continued to cry, "No, they are not coming, the gate has two locks on it and one can only be unlocked from this side if locked from this side. I saw one of the police running back to probably get help to open the door or to go to the only other gate up on the topside. We must hurry now to get out before they get there. I am fine; I was not hurt it is uphill from here, so let's go. I will tell you what happened as we go."

They ran through the tunnel all they could, but most of it turned to stairs about every twenty feet. Jill could think of only one thing now—that was hoping the police and other authorities would not go to the Temple of Hercules. She was still in shock as to what else happened. After a long uphill struggle, they came to the outside gate. No lights could be seen through the door except the dim streetlights. Jill and Sam shut their headlamps off and waited for Latif to unlock the door. When they got outside, they could see taxis off in the distant.

Jill turned to Latif and said, "Latif, we thank you sooo much. But we must get to the airport now. We must leave you here. I am sorry for that. But, but now I am sorry for your father. Latif!—I have caused the death of your father. The police were shooting at me, but your father stepped in the way and was trying to stop him from shooting. Latif, I am so sorry for all of this. I should never have come here to Petra. I am sorry to for making Sam come here with me and all this mess. Please, Latif, forgive me, your father told me about your mother. He told me to work with Raul to help you and Haifa find your mother's tomb. I don't know how, but I will. Will you want me to help you now? I will return if you want me to, but we must now go, please forgive us."

Sam grabbed Latif's hand and also apologized as he nervously looked down into the dark tunnel and at the roads leading into the parking lot. He then quickly grabbing Jill's hand and pulling her toward the taxi location. As Jill looked back at Latif for a reply of anger and distress, she could only see his glitter becoming a dull glow.

In a soft voice Latif said, "I forgive you both. Now my father is with my mother. My sister and I will find a way. I wish you well and a soon return to your family. I will be fine here; I hope to see you again someday. Now I am sad of what has happened to my father, on top of what we were planning to do. This has been the second most difficult time in my—life."

Latif floated over the wall and into darkness.

Sam and Jill walked in the street shadows to the nearest taxi and were on their way to the airport, which was about twenty minutes away. As they walked to a taxi, Jill turned, looking toward Latif. She put her hand over her mouth while tears again streamed down her face as she shook her head.

They met no resistance but had to go through the tedious steps before getting to the boarding room. When they went through the customs section, they were required to empty out their backpacks and almost strip down naked. Traveling as they did, the majority of items were cloths, most of which were dirty, sleeping bags, and hygiene items, with few souvenirs. The customs process did not take long, and they soon found themselves waiting for the boarding call. They ate most of their sack lunches, a roll of TUMS, and a couple glasses of wine before their boarding call came. As they waited in suspense, they continuously scanned the airports entrance door for trouble. Their red-eye flight time came with no other confrontations to deal with. They boarded the aircraft and were soon in the air headed for Tel-Aviv, Israel.

Chapter 6
Return Home

At the Tel-Aviv airport, they purchased tickets to continue their journey home. They spent the night sleeping on the floor of the boarding room waiting for their flight out to Amsterdam and then on to New York. From there, Sam was flying to Cambridge, Massachusetts, where he needed to report into his graduate studies' advisor and boss. Jill was going home to see her father before returning to University of Pennsylvania in a couple weeks to start her first semester of graduate school. They did not want to stray far away from boarding gates, so they choose not to find a hostel to spend a day and night at. From the waiting room, they could see the airplanes and that is all they wanted to see. They ate most of the food that was prepared for them and occupied their time with hesitant talk. They needed to come to a compromise on what they were to say about their adventures, other than just seeing some old archeological sites, eating many different foods, and meeting many different people. After spending about five days in Jordan, they could hardly remember the other four weeks of backpack travel.

As they lay up against the wall, Jill had her arms wrapped around Sam's neck. As she kissed him on the side of the mouth and stared into the side of his tired face, she said, "Sam, my dearest friend and love, I say we do not tell anyone, including family, about the ghosts. I think we should end it here, except with each other. I do think we should continue trying to figure out what all went on, but nobody would believe us if we told them, and we have no evidence. Not even one picture. They have done well eluding people for two thousand years, so I think they will be able to continue unknown. Do you agree we stay silent?"

Staring off into nowhere, Sam finished his pita bread and cheese and said, "I agree. I will not say a word, but I will study what I have seen. We also must plan to answer any questions about the death of Youssef—if such is ever brought out. I have a feeling that what all happened is going to be brushed under the table there in Amman. No matter what comes about, we never really did anything wrong nor is there much evidence pointing to us for anything. However, I will wrack my brain and try to find an explanation for this real *non*—paranormal adventure. It is so crazy. I think you should start a new degree. You could call it '*anthroghostology*'—wow, would the scholars of anthropology and archeology like that."

Jill wrapped her arms tightly around Sam's neck and said, "Right my dear. I would then be right up there on the podium with you and your '*soulology*,' or is it 'electron soulology'? Let's get serious, Sam, we have got to live with this adventure. It is something that will be with us forever."

Sam smiled as he slid out of her loosened arms and put his hands behind his head staring at the ceiling. He said, "Oh, I am serious, Jill; I think that we can secretly turn this into a really neat study on our own. Now just lay back and think about getting home and how neat this fall semester will be." Sam closed his eyes.

Jill flung her head back and stared at the ceiling as her long wavy blond hair spread over the backpacks. She slowly shook her head back and forth, as she tightened her lips. She then got up and wandered around the waiting room. She walked over to big bay windows allowing one to see the lights of Tel-Aviv. Reaching into her Colombia, zipper shirt pocket and retrieved the small scrap of paper, she got from Youssef. She gently unfolded it and quietly read aloud to herself:

She sleeps at rest unlike the best
Enter the grounds of six
There beneath is where one picks
The common stone as shown
Oh, be careful not to groan.
Then go below where water did flow
Go south until you are up in the house
Follow the squares of the golden snail
Until you reach its tail
There below you find your mother's hair
Which became entangled in such a snare.

Look in the color of gold
Just before that which you once did hold.

She read it again and again. It did not make much sense to her. But then, several things occurring over the past few days were not exactly clear. However, that was no reason to give up, either. She did have a good idea what the paper was—a map. She folded it up and put it back in her pocket as she stared out the window looking through the city lights to the south … toward Petra. She turned and went back to Sam, who was sound asleep.

Getting tickets for their next flight out at the airport did not give them a chance to sit next to each other the entire trip home. When they arrived at La Guardia Airport, they had a few minutes to spend with each other before they went their separate ways.

Jill's father wanted her to come home and spend at least one week there before she returned to school.

Sam was to report to Professor Steinhart as soon as possible so that he could get underway working in the lab as a graduate assistant.

After a long night alone and scattering his backpack contents throughout his singles' dormitory room, Sam headed off to Professor Steinhart's office. They had a long meeting discussing Sam's job and future classes. Sam had taken several undergraduate classes and a couple of graduate-level classes under Steinhart already, so they knew each other. By the time, Sam was in his third year of collage he had already taken about half of his graduate level required classes. He was mathematically gifted but was not interested in just numbers and equations. Sam's doctorate degree would read biophysics—this covered an extremely broad area of studies. Sam's favorite was that of nanotechnology, with the use of the up-and-coming science of artificial intelligence. He was particularly good in this area, which was one of the reasons he had been selected for the graduate position. His father was a high-school math teacher, and his mother taught biology at the small-town collage. Sam had grown up in science surroundings, which may have led to his accelerated advancement in science. He was also a very down-to-earth guy who liked spending a weekend just camping or snow skiing. In the world of nanotechnology, things were small—as in the size of atoms and molecules. For Sam that was not small enough. He was an admirer of the new group of scientist chasing electrons and buckyball electrons. He just wanted to steer away from the use of viruses and plain carbon sources to carry information and act as the smallest computers in the universe.

This is what he found interesting about the ghosts. He did not question whether they existed. He had no proof that they existed that he could show to the world. However, there was a time when thinkers could not prove to the world that it was round and not flat as taught. Brave ones had to step out beyond social and moral rules to prove a hidden or unwanted finding. He had made the decision to step out.

He and Jill had been told by Saud that they did not want to be flaunted in front of the world. They had a means to prevent that by simply not being actively visible before people. They had the ability to just be invisible beings who could not be seen or heard by anybody except another ghost. If the spice they used as a source of medium were to be put through the scientific method tests, it would just show spice matter and an energy source, which everything on and off earth had. That's what set Sam's mind buzzing. That is where the nanotechnology came in. Sam believed he had an even better way to see the ghosts. However, before he experimented with his hypothesis, he needed to go home also. It was just two weeks until school started, and he got things in order with Steinhart, whose main objective was to prove that matter could become self-replicating as a life form. He and others had already proven that molecules could be made that would copy themselves and form small strings of material. However, many did not consider these alive. They were referred to as artificial life. This completely phenomenal race had been going on now for several years. Most every person has seen a show or book relating to good old Doctor Frankenstein. However, today, there are many individuals, companies, and schools racing each other to create life or things that grow and reproduce on their own. This phenomenon is beyond science fiction today, and is a reality soon to be achieved, only in a matter of time, even as the anti-scientists battle them with words. Bio-laboratories are churning out results like the computer companies are churning out smaller and more powerful electronic boxes every six months. Professor Steinhart's laboratory was one of many that contained several projects of what are sometimes called "Primordial Soup brews." These are sterile conglomerations of various mixtures of amino acids, carbohydrates, nucleic acids, fatty acids, and so on—all of the building blocks of life waiting for something to cause a spontaneous generation to create an artificial life. Sam had been a fan of the two early scientists of the 1950s who took methane, ammonia, water, and hydrogen, and then added electricity and created amino acids. It was said back then that it could not be done by the non-creationist. Today, all the amino acids in life forms can be made. Those same types of close or limited-minded leaders said that Cristobal Colon could not sail across the ocean. To create a living individual was a bit different from an amino acid. The lightning bolt of life was what was missing to start a new life. Sam was not sure what he would produce in his works or exactly what he controversy he might create. However, he was determined to not simply fall into following the old book way or the thousands of myths taught and just follow along. He wanted to chase dreams his father talked to him about. So, he headed for home.

Jill had been home for several days now as Sam departed school. Her home was not quite, like it used to be. The loss of her mother not only changed her life but also changed her fathers.

He had been a very prominent cardiologist. He descended from a family of doctors. His wife was from a family of doctors. He was working to save a man's life in the emergency room the day his wife also was brought in. He was not aware she was in the same building dying as he struggled to save another's life. He succeeded in saving the old man's life yet failed in saving the one and only love of his life, with whom he had created another love of his life. Now he had only one real love left, Jill. He no longer worked in the field of medicine or health care. He diverted his medical and mental skills into helping people deal with the stressors of life. He needed no further schooling or degrees. He left the large cardiology clinic to open his own small but well-organized clinic of consultation and medicine. He became a somewhat holistic consultant. He no longer wanted to work in or around a hospital. The day Jill arrived home. Like Sam, she dumped out her backpack, but not in her bedroom. She dumped hers out in the laundry room. There she could sort out the few trinkets she had brought home with the dirty cloths and camping gear.

As she sorted through things, her Dad leaning in the doorframe and said, "I am so glad to see you. I have thought of you as well as worried about you each day you have been gone. I am glad you have gotten this out of your system and now can get to work on your graduate school. However, you already know that I will be asking if you have changed your mind on your graduate field. I still think you would love the field I am working in, and it is close to what your mother really loved: biology. Also, this work does have an income"

Jill interrupted. "Yes, Dad, I know I am going to starve to death as an anthropologist. Nevertheless, I love studying people. Mom did also—she even told me that the only reason she worked in social work was because it paid, but if she could, she would live near and study intriguing places like Machu Picchu, Peru. It is good to be home, and I am ready for school to start, and yes, I am going on with anthropology. Now, I am going to take a great long home shower and put something nice on so we can go out to our favorite restaurant, so give me an hour."

Her dad gave her a big smile and hug. "Okay, my dear. I will be in the den. Just leave your mess on the floor there and the house lady will wash it all. Each time I see you return from camping; I understand why I never did it and believe I will stay that way."

As he walked out of the room, Jill said, "I think I'll take you with me next time I go somewhere. That way, you can give me some good counseling." When he turned to look at her, she said, "Dad, I love you, and I missed you so bad sometimes I never want to leave you."

"I love you also greater than any trip or thing. And I will go with you on a trip as long as there is a hotel nearby for me." He winked and said, "Deal?"

"Deal," Jill replied as she went back to look around at the stuff on the floor. She piled her clothes in one place and camping gear in another, and then went upstairs to get ready for dinner.

They went to their favorite place called Texas de Brazil. It was a unique steakhouse or *churrascaria,* where servers walked around carrying large swordlike skewers of various cuts and types of meat. They place a little disk by your plate; on one side is written *Yes, Please,* the other side reads, *No Thank You.* While Jill's dad enjoyed the all-you-can-eat of fine cuts, Jill enjoyed the great fancy salad and side bar of cheeses, breads, vegetables, and other side dishes.

After being seated and served a glass of wine, Jill had to ask, "Dad, do you believe in ghosts or things like that?"

Her dad laughed, looked around for a second or two, and said, "My dear Jill, I can't believe such a question coming from you like that. You have always only believed in anything that you could see, feel, hear, taste, or smell. You are a realist and only a realist."

Jill quickly said, "Yes, Dad, I know I am hardheaded. However, do you think some of those ghost stories you hear about are true? I feel that most of them are false or simply a bullshit scheme made up to make some money on, but there are some remarkably interesting ones out there also. You have been around many people who you have seen die. What happens to them? I know the readiest answer to that question is 'Oh, they go to heaven or hell.' Well, today most have evolved out of the hell thing and just say heaven. That is, in our country, but there are many cheap answers out there like that. But, Dad, what if our being has both a physical and a nonphysical existence? I mean, can our soul really exist in a material manner that we have not seen, yet?"

Her dad put his fork down, took a long sip of wine, and said, "Jill! Just where have you been?" with a bit of a laugh. "Wow, well, you're right there, most people throughout the world believe in the myths, such as all religions. These they were taught from a child or forced to accept, just like why most—not everyone, but most people in Dallas are Dallas Cowboy fans. Just like most people in the U.S. are Christians because they killed most all of the non-Christians here and forced it upon the survivors. Moreover, as all religions are a needed physiological thing, we as humans need to put that fear aside. However, as the world gets smaller, that is slowly changing, as all things change. Churches argue and deface the education of evolution yet are the fastest evolving event in the human social medium because as our social technology changes, we must change religions to keep up in the market. However, as for ghosts, well, in general, I would say a ghost is what most people refer to as the spirit, soul, or mind when a person becomes deceased. The person's body simply becomes material stuff that would be good in the garden, but the real person, the soul, does continue to exist. Where it goes and what it does, I do not know. Then there are those things that are not a living thing or person that are also called ghosts, such as ghost town, ghost ships, ghost mountains, on and on. That is an area there is no complete answer to or proof of. I think it is just a psychological entity that we get, with a mind that has the ability to think or reason. We humans invent stories and myths to answer questions we cannot answer. Therefore, a ghost can be an imagination of someone who makes noise or moves things that we cannot explain, but they really do not exist. I think if they existed as people or things from other planets, they would just pop out in front of us all, making themselves well known. These ghost stories are myths just like Santa Claus, fairytales, and all religions. As you know, your mother and I were once loyal to and actively participated in the local church. As you are also aware, I gave that all up. I then became involved with the New Age era of spiritual seekers, the group of non-traditionalist teachers of the mind, body, and spirit. I went to many conferences and workshops that solved everything. They got most seekers all wrapped up in the mastery and leadership of the body, mind, and soul. They taught classes in integral life practices, which were to balance you and everyone else out in the world to become one, one with ourselves and one with God, whoever that might be. Oh my God, those workshops one sees in such magazines as *Enlighten Next* can solve all your problems. However, a credit card is very vital to those teachers, as with another church. They all have the answer for you and all charge well. In addition, if you are not healed or enlightened, the answer is 'Oh, well I guess it just was not your time." Or 'Oh, you just did not let your ego go. You did not surrender your control, allowing your compassion to form peace in your inner body.' My dear, many of the spirit teachers and religions try to sell one on the idea that one must turn the other cheek or maintain peace within oneself as well as outside of one. This I daresay is not the truth. This will not get your business ahead of Walmart. This peace will not save the sea lion from the killer whale. The real truth is to learn the laws of nature and not run from them. If you don't want to be bitten or eaten by a shark, well then stay out of his house. If someone slaps your face, turn your cheek and at the same time come across with a right hook. Okay. Enough rambling, let's eat."

Jill's face lit up as a great smile spread across it and said. "Dad sometimes you amaze me, sometimes you can ramble on a lot, but what you say is true, and I do accept your thoughts—except one. That is, I think ghosts do exist, but I, as all others, do not have proof. So, Sam and I are going to make chasing ghost stories a new hobby, so don't get alarmed if you see me reading something on ghosts or something like that."

Jill's dad laughed and said, "Okay, I won't. How are you and Sam doing? Anything serious yet?"

With a big smile on her face, Jill answered, "No, nothing like you might be thinking. We had a great time during the month we were exploring. It was so neat over in the Middle East. I think I have found the place I would like to do my doctoral thesis at. That is Petra, Jordan. However, I will have to do some politicking to get that opportunity from my school advisory board. There is already a crew working there, and they do not appear to want any other open-minded competition, which has become an anthropology and archeology epidemic. Most researchers out there do not want to share their space or findings for others to study."

Her dad joined in, adding, "Yes, that is a growing battle in most all areas where something new can be discovered or a profit made. Also, even now the education system does not want someone to rock the boat. So, when will you see Sam again?"

"I will stop by MIT before I go back to school. He is going home this weekend for a few days and then back before next weekend, so I guess I will see him for a day before we hit the books. Maybe I can get him to come here for a few days at Thanksgiving or Christmas break. We'll see."

Jill spent the rest of the week at home and then caught up with Sam just after he returned from his family visit. She met Sam at the MIT campus student union. They had a cup of Starbucks Caffé Mocha and talked about their visits home.

After a few minutes, Jill said, "Okay, Sam, show me your new home. I want to see your graduate room so I can compare it with mine. The building you are in looks like it was built back during the Civil War. I mean, the outside is old red-and-white brick, but the pictures you sent me of your room were really cool looking. I want to see the inside. Come on and show me." Jill grabbed her mocha and Sam's hand to pull him up out of his chair.

Sam said, "Wouldn't you rather see the lab? I mean, that's where I really live. I have my own desk and booth. I love that place. You will, too. How much time do you have before you need to take off for Penn?"

Jill let Sam's hand go but walked toward his dormitory building. "You are so close to the Charles River. You *are* going to take me scull boating, right?"

Sam slowly caught up to Jill and said, "Yes, I would love to. Thanksgiving break would be a good time for you to come by. We can go then. You know, before we get to my dorm, we will be going by the Plasma Science and Fusion Center. You wanna go see that?"

"No, Sam, I want to *huchecuche*," Jill said with a big grin and giggle. "I want to see your perfect room. I bet your underwear is neatly folded and on the left side of the second drawer. I bet your pants are hanging on the left side of your closet, and your shirts are on the right side."

Sam grinned a bit and said, "Yes, you are most likely right. Did you eat before you got here? If not, we could go get something to eat."

"Yes, I ate before I got here, and you know what I said I wanted to do," Jill said as she grabbed Sam's hand again and pulled him along.

They finally reached the graduate dorm on Albany Street. Sam's room was on the second floor. When they entered Sam's room, Jill slowly walked around with her hands clasped behind her back, acting like an inspecting officer. Sam's room was very orderly. There was a small blue couch with its back to the large windows showing the back patio area. There was a small desk with two laptop computers on it and three bookshelves above the desk fully loaded with books. Out from the couch was a small coffee table and then Sam's twin-size bed. It was made fit as a military bed. Jill walked over to the bed, turned around, and looked at Sam with a grin spreading across her face sitting in the one lounge chair next to his desk. Jill brought her hands out in front of her as she smiled at Sam. She then swung her arms back and forth, along her sides preparing to jump backward onto Sam's neat bed.

As she jumped, Sam slowly shook his head back and forth, saying, "You just had to do that, didn't you? You just cannot stand to see my bed made up because you cannot make yours. I bet your camping gear is still on the floor at your house, isn't it?"

"No, I think it has been washed and put up by now. I see you took all of yours home." Jill stopped looking at Sam. Her smile disappeared and then her eyes began rolling around slowly as if looking for something yet not seeing or focusing on anything.

Sam got up from the lounge chair, walked over to the window, pulled the curtains, and then smiled at Jill and said, "Okay, you wanted to *huchecuche.* He bounced from the window then stretched out diving forward for Jill on the bed.

She sat straight up, turned her head to Sam, and gasped, "You didn't!" then jumped up and ran from the bed all at the same time. Just after a couple quick steps from the bed, she turned around and said, "You did, didn't you?"

Sam looked at her without saying a word but tried to smile as he lay the length of the bed on his stomach. He rolled over onto his back, rubbing his palms together as a forced grin spread across his face, "Well, now that you have wrinkled my bed, I guess we can mess it up."

"You did! You did! You brought somebody home with you, didn't you?" Jill slowly walked back toward the bed. When she was about three feet away, she lowered herself to a kneeling position with her hand in a praying position under her chin. As she lowered herself to the floor, she kept her eyes on Sam, who turned his eyes to the ceiling.

Finally, Sam said, "Jill, wait. Let's talk. I will explain it to you, and you will understand. I really did not have time to explain it to you in Jordan. I was not sure you would understand anyway. But I think I can help them."

Jill bent over and looked under the bed, and then she slowly reached under it. She grabbed a small Tupperware dish and pulled it out. She slid it out in front of her and then gently pried the lid off and said, "A cake, a crazy cake. You put her in a cake and brought her home. God, Sam, just what in the hell are you going to do now? Are you going to eat it? What if it, or she, had been confiscated by the airport authorities? Someone else could have eaten her. Is she just going to pop out of the cake here in front of me or what?"

Sam sat upward flinging his legs over the side of the bed and reached out for Jill. His arms were not out for a hug of forgiveness. He reached his hand out for her to give him the dish and told her, "Imad's aunt, which is her aunt, also did it for me. Haifa told her and me that she wanted to come to the U.S. with me and that I could try using the methods I talked to her about to try and bring her back as a human being. Her aunt mixed her into the cake and then just heated it slowly to dry it. Now I just need to rehydrate the cake, separate the spice out, and let it dry. Haifa can then appear as a ghost. This is a practice Youssef's family has done before to help hide the ghosts in their house. You were gone to the museum, and this was the only chance we had to hide her."

Jill got up and walked toward the door. Just as she got there, she turned and said, "God, Sam, we get out of one mess and you get us into another. What you are going to do or try to do is wrong. Why did you not ask me first? Remember, we agreed to talk things over before we did anything."

"If I had asked you, would you have agreed?"

"Probably, no."

"There is nothing wrong with what I am going to do. Besides, Haifa asked me to try. You are the one who got us into this mess. I am just trying to solve it. Come back here next weekend and ask her yourself."

"It is wrong Sam! You are experimenting with life. You are playing a game outside of nature. What happened to them was an act of nature, but not just physical nature you're dealing with the life of nature, and I think you should leave it alone."

Sam put Haifa on his desk and sat down. They looked at each other in silence for a moment. Then Sam spoke quietly. "Look, Jill. I am just trying to help Haifa, and what happened to them was an act of man, humans acting on each other causing an unusual event. It's not just an act of nature or God as some would say. Now you sound like a narrow-minded creationist. Human beings unknowingly caused something to happen that I feel can be reversed. There are many laboratories around the world trying to create life or replicate cells. There are several awfully close, and one of those labs is right here. I think they are missing one thing, and that is the energy of life. I think …"

Jill took a step forward and said, "Sam, I don't want to hear about your great theories and thoughts now. Do you care about the ghosts of Petra's souls? Why not just leave them alone? Do you know who Haifa really is? Did she tell you about her life?"

Sam interrupted. "No! And I did not ask. That is all irrelevant. I am going to try to bring her back to human life here in the lab. You yourself want to talk with them and pry information from them about the history of Petra. Come on, isn't that just cheating in anthropology discoveries?"

Jill shook her head, turned, and walked to the door. As she went out, she said, "Bye, Sam! I wish you good luck. You'd better call Doctor Frankenstein." She went out the door and down the hall in a fast walk.

Sam got up slowly and went out to the hall railing. He looked down and said, "Do you want me to call you when Haifa appears or what? Is this the end of our relationship?" Sam got no answer.

Jill went straight to her BMW and drove to University of Pennsylvania. She checked in at her graduate housing office, unpacked, and arranged her room and things to begin her graduate studies. She was glad to see the weekend come to an end and start classes. This would get her mind off the ghostly encounter and on to reality, she thought.

However, she was not in class but for a couple of weeks when she bumped heads with another professor. This was in her graduate class of medical anthropology. The class had a group final exam that started the first day of class. It consisted of a project to piece together the remains of 158 human skulls that had been excavated in a temple mound site in Illinois. The skulls were found buried in small clusters of five to eleven. A few had the first and second vertebrae. By most of the archeologists on the excavation, it was thought to be a group of indigenous people who had been killed in warfare by another group. According to all of the measuring and methodology, it was determined that they were all men. However, in the final report that Jill's professor wrote at her lab, they were not all designated as men. Her professor informed the class that there was a big error in the field of science—that is, there was still a bit of male dominance within the science world wanting to jump to conclusions too much. She informed the class that based upon her findings; there were several females within the group.

When she brought this explanation out in class, Jill just had to ask, "But isn't that being a bit biased in your findings? After all, we cannot prove either way is correct, that is, all male or mostly male victims. Nevertheless, doesn't the methodology point to being all male? Then there is the cultural norm that most wars were carried out by male warring parties during that time period."

The professor put her hand out, stopped Jill, and said, "Jill, you must learn the disciplinary actions some professors take in this field and others. I cannot stand it when they try to gain all the credit for the discovery of something, based on just their interpretation of the facts. I was the one chosen to oversee and prepare the final report on this project; therefore, I must reevaluate everything and publish the correct findings. You, as a student, must watch me and learn the correct approach to the professional methodology. Okay, now, let's move on with the class."

Jill sat in her chair looking at her professor. With her mouth slightly open, she moved her eyes to the left and to the right, looking over the other students. All were quiet and soon began writing as the professor went on.

After classes that day, Jill sat at the Student Union sipping a cup of hot mocha and reading. She stopped reading to unzip the small pocket of her backpack. She pulled out the bit of paper Youssef had given her. She studied it for a moment, and then she took out some paper and started drawing. She then sat back in her chair and threw her pencil up on the paper. After a few minutes of stillness, she got her phone out, looked at it, pushed a few buttons, and put the phone to her ear.

Within a couple of rings, Sam's voice said, "Hello there, stranger, I see you haven't totally forgotten me. Have you forgiven me?"

Jill answered, "Yes, kind of, but I called to ask you a question." She then told Sam about the note Youssef had given her and read it to him. She told Sam that it was a code Youssef had written or somebody else had as to where Katrina was buried. She also told him what Youssef asked them to do. She asked him if he would help her try to figure it out.

Sam said, "Okay, send a copy of it to me and a copy of all the pictures you took in Petra. I will look those things over and see what I can come up with. What are you going to do if you learn where she might be and how are your classes going?"

Jill smiled as she answered, "Thanks for asking. My classes are fine. But I am ready to get out of the classroom already. If I can figure out what the code means, I will go back there to search for the location. I will do whatever I can to find Katrina because that is what Youssef asked."

"Is Katrina a ghost?" Sam eagerly asked.

"I don't know. He did not say. But I don't think she is, or we would have seen her. Youssef just asked that I find her tomb and ask Raul to help me. I think he is involved with this somehow. I think I can get any charges dropped on us if I help him to find her. That would give him some credit there or answer something for him. It would also get me some credit for finding something. Sam! He was interested in us as much as Youssef. He never turned us in to the police until he had to in Amman. Help me with the puzzle so I can call him to get permitted to go back there … and Sam, would you go with me?"

After a bit of silence, Sam said, "I don't know. I might. I will just have to wait and see how things go here. I will get back to you in a few days."

Jill asked, "Have you started your process? I mean, is she out of the bag?"

"Yes. I will call you in a couple of days, okay. I gotta go for now."

Jill pushed the end button on her phone and just stared at it for a moment then picked the paper back up and read it again. She then packed her things, went back to her room, and began downloading all the photos she took and sending them to Sam along with the code. As she looked over the photos, there were several that really popped out to her. They were the pictures she took of the old Byzantine Church and the Qsar Al-Bint Temple.

The side isles of the church floor were covered with various mosaics made from tiny pieces of colored stones. She found one picture that she believed must be a mosaic of the Qsar Al-Bint Temple. It showed the largest build that could have been in Petra and no doubt depicted nothing but royalty standing around its great columns and bench areas.

She wrote Sam a letter telling him to look at the photos of the church and the temple. At the end of the letter, she asked him again if he would go back there with her. She also asked how his project was going. In closing she said, "Sam, our romantic love may no longer be with us, but I still love you as the greatest friend other than my mom and dad. I did get mad at you for what you did. However, I think you did it more for Haifa than you did for yourself. It shows. It shows in that concerned look I see deep in your eyes. I see that look you have when you are in that, come hell or high water, state where you are going to do no matter what. I wish you well on your project, Oh and Sam please don't forget mine."

Sam retrieved all that Jill sent him. After reading her note to him he smiled big and began looking the photos over quickly and then turned to his project. He swiveled his chair toward the window and said, "Look at these pictures. I realize that this place is a bit of a culture shock to you, and you are a bit homesick and not quite sure you're happy here, but it will get better. Jill wants me to look over these pictures of Petra, especially four of the great temple and the old church. She also sent this code or poem-sounding thing of the place where Youssef told her Katrina was buried. She said Youssef wanted her to go find Katrina. Why, I do not know. Anyway, here we go again, jumping into this Youssef and Katrina thing. Listen to this and tell me if it rings a bell."

As Sam pinned the four printed pictures to his billboard Haifa's soft voice fill the air. "It's not a temple, it's more like a castle, but really it was a home."

"Sorry," Sam said, "It's hard to get what one has read from books out of your head sometimes. Okay now listen to this."

Sam read the Youssef lyric to Haifa, who came near the computer desk to look at the photos Sam had put up as well as the others on his computer. As Sam read the lyric aloud, Haifa moved backward until she bumped the wall at the head of the room's bed. She then descended to the floor in a crouch. When Sam quit reading, he snickered a bit and said, "What a story, a story to a grave. Now she wants to go find it to prove something to the archeologist there, and I guess some here. God, she is stubborn." He turned to the side a bit and asked, "Well, what do you think? Mean anything?"

Haifa was not standing nearby any longer. He turned further around to find her hovering with her knees drawn up to her chest and arms wrapped around them.

He quickly said. "Ah, what is wrong? Are you all right? You look … you look dim. What is wrong, did you hear the story I read?

Haifa remained motionless and then answered, "I am crying. I am sad."

Sam got up from his desk chair and went to the bed. He sat down near his pillow and looked down at her, asking, "Is it something I have said or the pictures? Ah, I bet they brought back old memories making you more homesick."

"Both," she said as she turned and looked to Sam.

"Okay, sorry now, how can I help you feel better?"

"Sam! Katrina was my mother. Youssef was my father, and Latif, Latif is my twin brother." Sam looked down at Haifa a moment and then up at the ceiling. He took a deep breath and fell backward on his bed, covered his eyes with his hands, and said, "Oh, shit. This is not going to work. What have I gotten myself into now, again?" Haifa gently rose up and moved over near Sam's waist with her knees folded under her. She moved her hand out and placed it over Sam's heart. Sam quit rocking his head back and forth.

A dreamlike smile appeared across his face as Haifa said, "Sam! Nothing has changed. I still want you to take me to your laboratory and perform the operation that will bring me back to human-being life. I want to be reborn. I am sorry for not telling you about my past. There is so much to tell you, but it will take days, weeks maybe, years to tell you all of my times. I am sorry for what my father tried to do to you and Jill. He had no other idea of how to help us become reborn like he was. He …"

Sam uncovered his eyes, stared at the ceiling and said, "Wait, wait, let me get this straight. You and Latif are brother and sister, okay, I understand. Katrina the king's daughter back two thousand years ago was your mother. Just how in the world is Youssef your father?"

Haifa sat still on the bed, leaving no impression on it but looking as if she was comfortably sitting on the sides of her calves. She told Sam about her family's past life during the peak of the Nabataean empire and the takeover by the Romans. She moved her hand around over his chest and continued saying, "I can feel your heartbeat. It was racing, but now it is slowing down. Youssef was my father back in the real world I lived in before we became ghosts. He, my brother Youssef, and I all became misplaced souls, but then just about forty years ago our father became reborn as a human being. How!—well, kind of like some of the old Romans. If my father and brother know, they have not told me. However, my father told us he knew how to make us alive again or reborn again into the human-being form. When the time came, and I was told what I was to do, I refused to. I disobeyed my father."

Sam sat up as Haifa's hand moved around him and then asked, "So Jill knew about this plan? She did not say a word about it to me. How long did she know about it?" Sam reached into his pocket, retrieved his BlackBerry phone, and said, "She knew all along, yet gives me shit about what I am doing. I need to call her and tell her thanks for keeping me posted."

Haifa moved her hand up to Sam's arm as if grabbing it just before his hand. As she did, Sam released the phone, which bounced off the bed to the floor. As it hit the floor between his feet, Sam grabbed his hand and yelled, "Dang, stop that, wow, what the heck was that? It felt worse than sticking a bobby pin in an electrical socket. So, you can move your energy or power. You have control over where it goes, don't, you? Youssef did this to become a new person. So, it will work."

"Yes, I have some control of my energy, but it makes us weak when we do things like that. I can make you feel tingly or shocked, but don't worry, I cannot kill you. By touching you, I can feel your feelings. I can almost read your mind. However, feeling one is better than reading one's mind. What people say is not always true. But, to feel one, oh Sam, it is so, so … so good. And your feelings are even better than good. When I am reborn, I hope to teach people how to use their soul while being just human. Start your project with me, and when you have started the process to bring me back, listen to your songs you like. Remember me when you listen to song number two and four on you U2 album."

Haifa slowly moved from Sam's bed and went to her new Tupperware bed. She told Sam that she was tired and needed to rest. Sam reached down and picked up his phone. He looked at it and pushed the roller ball to activate it. It was dead.

Haifa looked at Sam and said, "It's just out of power. I borrowed some, now I go night-night. So, wake me when you are ready."

As she slowly sank into the dish, Sam walked over and got the lid off his desk. He fastened the lid to keep from spilling Haifa and put her in his backpack. He gathered up some paperwork and his books, and then he went to the lab.

When he entered the lab, Professor Steinhart called him into his office. Sam laid his backpack on his desk chair and pushed it under his desk. He then went into the professor's office and took a seat as Steinhart continued writing in a notebook.

As he wrote, he said, "I am so glad to see people like you attracted to and supporting such great concepts growing out of science today. Unlike such narrow-minded individuals as this one, I am writing to in confrontation of my work. No doubt, he will be at the conference I am headed to. As much as we are discovering today, the only thing most people are interested in is a quick fix to their obesity, wrinkles, all diseases, and short life. There are several of us who are just so close to recreating—or I should say creating—a fully self-replicating form of life."

He looked over at Sam and then went back to writing and talking. "When one of us does make the big break, though, there is no telling what hell we will receive for simply making an artificial organism."

Sam interrupted, asking, "Dr. Steinhart! Why would you call it an artificial organism? Would …"

"Oh, my dear friend, I did not give the outcome that name. That is something others, mainly the anti-scientists and creationists, have done. They see the outcomes of science growing so fast their old religions cannot keep up. Nevertheless, it will be human made, so I guess that is where that term comes from. As far as I am concerned, it will be the reinvention of living cells set into motion of reproducing itself and changing its physiology to fit its environment just as all life forms have been doing for eons here on earth and maybe elsewhere."

Professor Steinhart stopped talking, put his pencil down, and stood up. As he walked out the door he said, "Okay, young man, let's take a walk around the lab and make sure you understand all that you need to do while I am gone for the next week. Also remember, I need a ride to the airport tomorrow morning—be here at eight. Now, we have gone over everything in here that needs attention, as well as who can come and go here. Always keep the doors locked remember if someone has access into this lab then they will have their door code. Other than that, no one is allowed, friends or not. Now over here in this second locked room where the a-life incubators are, remember, you must check these incubation tanks three times a day, no if's, and's, or but's. That is, at seven in the morning, at two in the afternoon and ten at night. I do not want them going more than ten hours without being checked. These projects are nothing new to you, and I hope to see you and your nanotechnology endeavors take them over someday."

They slowly walked around the room as the professor checked the computer data station at each experiment setup.

Sam said. "Yes sir, I will give each and every one my undivided attention. That includes the starting of my two crystal growth projects that I talked to you about a week ago."

"Oh, yes, yes, that will be fine, you are welcome to the one unused incubator, and for your other one, you will just have to ask around at different departments to borrow one. Make sure you get a signed contract so no one can throw your work out to fertilize the grass."

The next morning, Sam drove Professor Steinhart to the airport. On his way back, he called Jill and asked her if she could come visit him. He told her he had some good news for her and that he also wanted her to bring the theme paper she had been researching on the crystallization versus sandstone that was going on in Petra to determine the disintegration of the building facades and other structures.

She agreed to visit and said she would be there Friday.

For the next couple of days, Sam carried out his daily routine of classes, maintaining the lab, and gathering what he believed would be the building blocks of a reborn human being. Now to most people, if not all, this was absolutely impossible or ludicrous at the least. Sam did not have the total knowledge or supplies to stir up the batch of life-supporting "primordial mud," as some so called it. But he did have access to several choices that had already been made up. Of the several incubators in the lab, there was one that contained the perfect mixture of every element of all humans as well as all other living organisms on earth. It also had the right amount, that is, about 120 pounds. He could not use the lab incubator for fear of the doctor getting interested in it or changing it. Therefore, he developed a plan for a leak in the one he wanted. He then made an apparatus to collect the nutrients from that catastrophic accident so that he could transport the required elements to build a human structure in a sterile way. Sam felt that several scientists were on the right track but were simply missing the main ingredient and that was the catalyst of life. Sam had that part but also did not want to share.

When Jill arrived on Friday, they met over a cup of mocha. Sam discussed with her his plan and that he needed Jill to visit the director of botany in order to obtain permission to use this special incubator in his experiment. That way, it would be out of Dr. Steinhart's sight and out of everyone else's mind. Jill agreed to go through with the plan without listening much and then asked.

"Sam! The code, what does it mean? You said you had some good news?"

"Well, yes, I have looked at the pictures and think there is a way we can find out what they mean. However, let me tell you this. For years, scientists have always thought life forms have only used part of our DNA and RNA. About ninety percent of it for years was thought of as junk RNA/DNA, but over the past couple of years, some have learned to use what is called "microRNA." That is what controls the building and repair of our organs and life. Well, really, the life is, I think, the missing ingredient in all of these projects. However, I have life. That is, the life that is in every soul and maintains the DNA/RNA code in the electron. After all, that is the way computers move and store codes. With Haifa, I have the basic catalysts to put the rebuilding of the body into effect, and …"

Jill put her hand over Sam's mouth and said, "Sam, Sam! Wait. That is enough. I believe you and support you. What does the code tell you?"

Sam smiled and replied, "Okay, I could go on forever with this. Okay, the DNA code …"

Jill interrupted again. "Please stop, Sam, do you not hear? No more. Don't tell me one more thing about DNA/RNA and your project. I want to know about the writings Youssef gave me. I will help you get the mummy tank, so let's go. But before I leave, I want to talk with Haifa."

Jill got up, grabbed her mocha, and headed for the door.

Sam got up and followed Jill to his room. There they spoke with Haifa but learned nothing about the location of her mother. Haifa told them the story again, about how she was told that the new Roman leader of Petra had taken her mother away. She and her brother were of no use to the Romans and were therefore imprisoned with their father in the spice storage vault. When they emerged as ghosts a hundreds of years later, they learned through listening and gathering information that the king and his family died in slavery, but this information was only secondhand. They wandered the city for years trying to find her, hopefully as a ghost. After hundreds of years, they were sure that all ghosts had been found, other than those that snuffed out a few Romans. It was certain that her mother simply had died, and her soul naturally ascended to the Otherworld as most all souls do, along with those of corruption.

After talking to Haifa, they went to requisition the incubator for Sam to use to regenerate Haifa into a human being.

During their walk, Jill finally broke the silence between them. "Sam, I do support your plan. I also have learned a bit about myself here lately after talking with my dad, you, and of course, the ghosts. All this confusion has led me to some soul-searching. I am an evolutionist, as you know. That is a bad word to some, but you know, it has just dawned on me that it is a science like many others if you read about it, wonder about it, or question it. However, you can go out and test it with the senses God gave us. God intended us to use our senses and our thoughts. There is no doubt in my mind that, God also is an evolutionist, just look at our constant changing world and you can see that. Religions are all fabricated, or as you say, human-made. None of them can be tested. The leaders and followers of all of them simple rely on one or two little fabricated books or stories. Sam, I think what you are doing is just exploring life. What you are doing does not affect life at all. I have learned that life is immortal, and it is simply in our souls, and that they, our souls, live in the same old world of matter and energy put together making us. Where the Otherworld is, I don't know. I will just have to wait to catch the aurora borealis to find out. Nevertheless, Sam, I have got to go back to Petra. Will you go with me and will you please figure out the lyrics of Youssef? I am going to call Raul in Petra and tell him I know the secret he is looking for and leave it at that. Then we can get his permission to go there. We need him to ward off the archeologist guru also."

Sam caught up with Jill and said he would work on the lyrics after he got his project going, which would be within the next week. They had their meeting with the botany director and got the permission to use the incubator after Jill explained her joint plan with Sam on crystallizing and de-crystallizing minerals along with sandstones to try to reconstruct what the archeologist were trying to prove at Petra. Jill was conducting the process on her own at University of Pennsylvania and was not going to be really doing it here with Sam, but it got Haifa out of sight and in production in Sam's hidden location. Before she left for The University of Pennsylvania, Jill went with Sam to his room to say goodbye to Haifa. She told Haifa to think positive in her dreams and hoped all would turn out well.

Jill said her goodbyes and started to leave when Haifa said, "Jill I think I understand your hard feelings in several ways, but I do thank you for helping me and my brother. I also thank you for trying to find my mother's burial site. As you know, that will help to give my brother and me a closure in this world. It will also help you in your field in proving someone else wrong. I also hope to be with you when you go there soon—so wait for me. Sam is not creating life. He is just creating a human structure for me to borrow for my time here to replace the original one I lost. I do not fear anything—well yes, there is one thing I do fear, but never mind that. It will not be an artificial life because if all works correctly as planned by Sam, I will be reborn in the world that I really never left. I will then be able to rejoin my father and mother as you will when my time here is complete."

After listening very carefully, Jill walked to Haifa and hugged her, and said, "You fear that you may not see Sam as human to human, right?"

"Yes, and I am sorry for evading your life, but I did not see the two of you as—well, as soul mates but more like, well just exploring mates as we all do as we grow and blossom out into life."

"Yes, Haifa your right. We are now really closer to each other than before. I will help him, then together we will go back to your home." Jill hugged Sam and left. After she left, Sam told Haifa that he needed to go to the lab and prepare the incubator. He reached for Haifa's hand. As she grasped his, he said, it would be ready for you on Sunday. Once we begin the process, I will not see you again until about the first week of December. That is based on the growth rate I have calculated"

Haifa glittered a bit more than usual and said, "I am ready. However, Sam, Jill feels she has lost you a bit, so you must repair that with her. As I am being reborn please help her with the code."

"She has not lost me. We will always be at least friends and share a very trying experience in our lives. We will talk to each other over the next couple of months, and I will tell you about it later, but right now, we have the most trying thing to do."

"Sam, Leave the music on for me while you are gone to prepare the lab, it kind of takes my mind off of everything."

"Okay, I will, but you must deactivate. Now come Sunday, when no one is around, I will take you to the lab. Then you can activate for the last time. All of your energy cells will be in the right location. I most likely will have to work through the night to get all ready, okay. But do not wear yourself out while I am gone"

"I will not wear myself out. Put my little bed on yours and I will descend into it. Leave your stereo on for me and fully loaded with our favorites." Haifa said as she waited at the side of Sam's bed. Sam came close to Haifa for one of many hugs they had now experienced but this time it looked as if Sam was entering into Haifa. He was preoccupied with the greatest endeavor of his life as well as that for another. Her then released her and walked to his desk, picked up the Tupperware dish and placed it in the middle of his bed. Haifa floated up over the dish and slowly descended into it as the scent of clove filled the room.

Sam went to his graduate office to spring the leak then gather the ingredients for Haifa. Maintaining the correct temperature was always important in these projects, and of course preventing contamination. For this, he had constructed a movable temperature-controlled apparatus to move the components from his Professors broken incubator to the botany lab. After getting the primordial mud ready to gather, he went to the botany center research greenhouse to prepare the loaned incubator. After an unknown number of hours to him, he discovered he was getting hungry. It was Saturday morning already. He had worked through the night without noticing the world around him, but the incubator was ready.

He went to the Student Union to eat and then returned to his room, where he collapsed on his bed exhausted from mental and physical stress. After he dozed off and sleeping for a few hours, his eyes popped open. He lifted himself up off the bed as if he were late for an especially important class or seminar. He grabbed his backpack, stuffed some paperwork in it along with the Tupperware dish, and headed for the lab. After doing his job at his lab, he headed for the greenhouse. Everything checked out. He made sure the doors were locked a second time, and then he set the dish out, opened the lid and waited as he made final preparations. He could not just sit and watch the dish, as well as he could not help but go over and over the computer data and the operation of the incubator. After about a five-minute data check, he went to the bathroom for the second time. He washed his sweaty hands again and then went back into the private lab room. He was stopped dead in his tracks by Haifa just outside the door. She said, "Sam, it is just Saturday afternoon. Is everything all right?"

Sam stood, nodding his head rapidly up and down, and said, "Yes, everything is all right. The incubator is ready … for you. I think we should start." Sam held his hand out for Haifa and then squinted his eyes shut as Haifa grabbed it. The graduate project room had a sign on the outside of the door with Sam's name and phone number on it as well as a "do not disturb" sign. Inside the room, Sam put his hand on the incubator lid, looked at Haifa, and said, "Well, here is your new home for the next couple of months. You should be ready to wake yourself about the first week of December. I will be here each day waiting on you. The case is locked from the outside. No one can tamper with the experiment. But don't …"

"Sam, I am not an experiment. I am me. I will wake myself with no problem when my new body is ready. I will be fine. However, if I do not form correctly, just stop things and somehow try to release my soul, maybe I will return as I am today, or I will see my Father and Mother. After all you have told me, I think I understand what will happen."

Sam smiled at Haifa and said, "All will work out. After all, we have got to go get your brother. So, think about that while you sleep. Also, I installed an iPod that will begin to play music each day at seven AM and then turn off at eight PM each night. That will get you used too day and night hours. We will work on walking after that."

Haifa moved around the small room, looked over her new little home, and said, "After all the scientific talks you gave me, the best way I can think of it all reminds me of sitting in a car one time as it began to rain. The car had just been washed and waxed. As the small slow raindrops fell out of the sky and landed on the hood, little pools began to grow. As one became larger from a few drops, it grew closer to another. When two of them got within a certain distance, all of a sudden, they just slammed into each other as if they were magnets. Soon after, several they formed a really large puddle. Then a stream formed that soon found another stream or another pool then off the hood into the earth the water ran. That is what is going to happen within me. My new elements are going to attract each other as my soul directs them. I just need for you to come and unlock me so I can then breathe and return to earth. I am ready. Do you want your ghostly kiss before I go?"

"What? What did you say?"

"Remember back in Petra when you were asking how we would feel together and all those other questions you had. I said, 'Do you want a ghostly kiss before I go?'"

"Is it going to shock me like when you made me drop my phone?"

"No, Sam, it won't shock you … but I cannot tell you really how it will feel because I have never kissed as a ghost. So, tell me how it feels."

Haifa floated within inches of Sam. Being a bit shorter, she had to look up into Sam's eyes. As Sam looked into Haifa's face, he could see her smooth facial features and into a sea of billions of little glittering stars. When his lips touched hers, there was no electric charge felt, but a shock of life was. Radiating warmth flowed through his lips into his whole body like being filled with a slow galvanized vibration. Not only did Sam feel a shimmer run through him, but Haifa did also as well. She began to glow like a crystal prism filling the room with tiny colored rainbows of internal light. When Haifa flowed back a few inches, there was a silence between them. Both were left speechless.

Sam unsealed the sterile incubator and raised the lid. Haifa leaned over and flowed into the readymade crib. She then just lay back as if in a tanning bed with a semi-liquid thick white blanket and began to sink into it like quicksand. Sam sealed and locked the lid. He then went through his checklist while going through the monitor data. Finally, he sat back in the one and only chair, leaned back, and closed his eyes.

Suddenly, Sam found himself falling out of his chair as much-needed sleep tried catching him. He gathered himself, checked the apparatus, and left the laboratory. When he got to his room, he sat on the end of his bed, pulled his phone out, rolled the little ball around, and located the contact list. When he found Jill's name, he started to push the roller ball, and then he rolled it again until he found his music. He looked at the two numbers Haifa told him to listen to. When he pushed the roller ball on the number two songs on the U2 album, "I Still Haven't Found What I'm Looking For," *filled* the air. He lay back onto the bed. It had been a long tiring day, night, and now day again. Sam tried to relax as his mind sank into the rhythm of the music. As usual, Sam's mind listened to the words. When the song ended, his eyes popped open. He quickly rolled to the next song number, which was number four. "With or Without You" played. Again, he relaxed and listened. The second time he played the songs, tears began to form little ponds over his eyelids and soon began flowing out over the side. The thriving beat of the drum soon got his heart racing. Suddenly he sat up, dried his face, tossed the BlackBerry on the bed, and went to his desk. He grabbed a poster off the wall, turned it around, and re-tacked it to the wall. Then, grabbing a marker from his desk drawer and began drawing. He drew a large rectangle with the long side going up and down on the posters back. Then he divided the rectangle into two parts, a large square taking up more that the bottom half, leaving a smaller rectangle in the top part. He did this several more times with each smaller rectangle that appeared. He stepped back a bit and studied the sequence of smaller rectangles, which always left a square. Then turned his laptop on and went to the pictures of Petra. He found the one he was looking for which was a picture of one of the mosaics inside of the old Byzantine church. He zoomed in on the floor or ground depicted in the mosaics. He began shaking his head up and down, turned to the bed, grabbed his phone, and called Jill.

When Jill answered, Sam said, "I got it. I think I have found what you are looking for. I am not exactly on it, but I am close. It has something to do with the famous *golden rectangle*"

"Hi, Sam," came Jill's voice. "How are you doing? I am fine. Okay, now that greetings are over, what have you got?"

"We need to go back to Petra. But, before we do, I need some more time, but I think I can figure this lyrics thing out. I will get back to you in a week or so."

"God, Sam, sometimes you are so difficult to talk with. What did you find out, and how did things go with starting your project? Sometimes I think you should get some anti-anxiety pills."

"All went well with the start. Now I have to wait for several weeks. I hate that. Okay, as for Petra. The grave of Katrina is going to be near or in the castles. I think you will have to do some excavating, so get going on talking to Raul. You are going to have to get permission to dig or move rocks. You're going to have to butt heads with that archeologist running the show there. I think this little note Youssef gave you will get you awfully close to Katrina."

"Why is it the castles, Sam? That is not where they buried people"

"The picture. Look at the picture of the castles or any picture of it in a book or on the Web. The ground out in front of it was covered with large six-sided stones. That's the grounds of six."

"What about the rest of the lyrics?"

"Well, I don't know right now, and I need to go to work in my lab. Call and get us in there, we are going to have to go back there to solve the other parts."

"Okay, Sam, I will get all that taken care of. Maybe we can go Christmas break. Oh, by the way how is Haifa?"

"Fine. All is well so far. I will call you in a week or so or write you. Bye for now."

Jill told Sam she would get to work on sweet-talking Raul and getting him to try to get permission to do an archeological exploration of the castles for a graduate project. She would wait for Sam's call with more news.

A week after his departure Professor Steinhart returned learning that one of his prize matter-forming incubators had developed a leak, draining all out on the floor and keeping Sam busy all night cleaning up. Sam had even brought the apparatus out to the cleaning station in the big lab, disassembled most of the parts, and cleaned it, readying it for the professor to start another.

Steinhart was not happy with the disaster but did not waste time in brooding over it. He quickly gave Sam instructions on refilling to tank with the quantum electrodynamics project they had been working on together. Sam was relieved and now had just one major project to worry about.

For the next few weeks, Jill stayed busy making connections with Raul. When she first made contact with the guard, he quickly expressed his anger toward Jill and her counterpart with a bit of revenge in his words. After airing himself, he asked why she had the nerve to call him after causing the death of an old friend and leaving him to explain what had happened.

Jill response was, "Youssef, your friend? Why is it he was always kind of running from you? Why is it you always wanted to know what he said to us? Why didn't you just ask your friend yourself? Anyway, all that aside, he told me to contact you for help and that there was something I could help you with."

"What did he tell you, and how is it you can help me after all the trouble you have caused?"

Jill sat silent for a bit and then said, "Youssef told me where something was, something especially important, and that you might be interested in it. I think it is something that would boost your job rating a little."

Raul quickly asked, "What did he tell you? Tell me what he said so that I can see if you are lying.

"He told me where I could find the king's burial, along with other things."

Raul was quite for a moment and then said, "That is not possible. Youssef, himself did not know where the king was buried."

"No, not exactly, but he gave me the code that gives the secret location."

Raul laughed aloud and then said, "Yes, yes, yes, so what if you have the code. ' She sleeps at rest unlike the best.' I also have heard this crazy set of words. That is all that it is, an old set of words. I think someone tried making a poem. How can it have anything to do with telling where something is that was put into the ground thousands of years ago?"

"What if the code does tell where the secret tomb is, and I know what is says."

"Where is the location? Tell me where you think it is, and I will tell you what I think."

Jill then laughed a bit and said, "No, no way. But I can show you where it is. However, I also want something. You must let us return to Petra without trouble and give us assurance that I get what I want when I get there. Do the Jordanian authorities want us for anything? What happened with the shooting of Youssef?"

"The shooting was an accident while the police chased a couple of robbers who broke into the museum. The trespassers escaped and have not been found to this day. They were recorded as local troublemakers. I had no other proof of who they were. I can send you the report that was put in the paper if you send me your address. What is it you want from the tomb?"

"I would like to have just one small item as a souvenir and a reminder that I led you and any other official to the tomb. This will give me credit for finding it to boost my career. I just want some credit in helping you find what you have been looking for."

Raul was silent for a long moment and then said, "I cannot let you take something from this country. Now, tell me what it is you want.

Maybe we can work something out. You are not wanted in this country for anything. The police do not know your names, and I did not report you to anyone. You are welcome to come here, and we will talk."

"Okay. We will be there during our Christmas break. We can only be there for two weeks, no more. Therefore, you must allow us to lead you to the spot, and then we dig. We do not want to have to wait for further papers or permission to continue once we have showed you the way so that you have it all to yourself when we have to leave."

"Very well, I will wait for you to call me again just before you arrive."

They finished talking over their agreement, and then Jill sat back in her chair and smiled to herself. She called Sam to inform him that they were going to Petra the first day of Christmas break. Sam agreed and then informed Jill that he was sure where the general, location of the tomb was, but that he would have to be there on the site and enter a place to find the correct starting spot within the remains of the castle floor, which would be when *you're up in the house.*

He explained to her that when one looked at the mosaics in the old church, it was obvious to see the grounds of six. This was the large yard out in front of the castles. There was about an acre of ground covered with large stones cut into hexagon shapes that tile the ground. Also, the mosaic showed the middle of this yard with a large pineapple drawn out on the ground. Each of the six-sided stones was a skin patch on the pineapple, just as one would find on a pineapple at any food store. There were various trees, fruits, and flowers surrounding the pineapple. However, the pattern on a pineapple had the hexagon skin patches in patterns of five, eight, and thirteen. This was the pattern of the *golden ratio,* which went along with the golden rectangle. These rows of skin patches ran diagonal across the sides of a pineapple, all with a common patch. That is, the diagonal lines made up of the six-sided skin patches of five, eight, and thirteen all had one patch they run through, *the common stone as shown,*--- the door. From the pictures, Sam could see there was a link, but could only see a few faint markings of the pineapple in the pictures if any. He said they needed to go to the castle yard and find where the layout of the pineapple was. Then they would need to raise the common stone in the yard. That would put them in an underground tunnel or old aqueduct leading to somewhere under the castles, which was once Haifa's and Latif's house. At the end of this tunnel, they would go up through the floor to determine where the head of the snail was. From there they would locate each square, as they got smaller, *until you reach its tail.* Throughout history, there have been many works of design made with the golden ratio of this rectangle. This rectangle, which Sam found to be that of the castle's floor from looking at the pictures, can be divided into a large square, a small square and one rectangle forever and ever. The mathematical ratio of the castle's floor squares are in the same sequence as the hexagons on the outside. These squares get smaller and smaller as they get closer to the tomb, at the snail's tail. That is where they would have to dig for the tomb. However, the head or starting point leading to the tail would be the largest square inside the castle, where the tunnel came up from the outside to the inside, which was the snails head. He could not determine the exact location of the king and queen's tomb—or where their daughter, Katrina, the mother of Haifa and Latif, was buried—without being at the castle. Finding this great talked-about tomb would give Jill credit for finding a great discovery by way of research and secret code-breaking. It would also give the Petra senior guard, Raul a boost in his career as being the lead investigator.

As for Sam, this tomb-finding thing was small potatoes. He wanted to do something a bit more controversial. He was breaking more rules of nature than Jill's project could come close to. He was defying the rules of human society as well as bending the laws of science to reestablish a life from a lost soul. He had determined the soul electrons contained the genetic code containing the code of life. His endeavors might seem out of reach to many and totally ludicrous to some, but it was something that was being worked on by many others. Most people would have told you the cloning of a sheep was impossible thirty years ago. Today there are hundreds of plants and animals being cloned for the purposes of science as well as for money. Sam was not cloning; he was simply letting nature take its course in reverse in a controlled man-made environment.

Jill would lie in her bed each night, having a hard time falling asleep. She could not rid herself of the guilt as to what happened to Youssef. What if it had been her father, who would also have been one to do almost anything to help his daughter? Classes were beginning to take a backseat in her life. She kept in contact with Sam almost daily on the computer but not by phone. Sam was not a big conversationalist when it came to the phone anyway, but he also was heavily occupied with thought each night. He wondered what he would do if his plan went wrong. What if it did not work? That could lead to several outcomes. Could Haifa literally die? Even her soul? Could she completely reform into a reborn human being but with mutations that were prominent like a missing arm or leg? Could she be reborn but with a different personality? What would he do if that happened? He had ways to monitor her progress, and if he saw something go astray, he could … abort the reforming process. Haifa would be like a child in an unfinished stage, yet still the same child. He had made a commitment within himself to allow Haifa to simply be reborn. If it totally failed, he would retrieve the soul electrons and go from there.

The short holiday school break for Thanksgiving soon came along, giving Sam and Jill time together. Jill went home to her father in Boston, and from there they drove over to MIT and picked up Sam. Sam and Jill needed to be together to discuss how they were going to Petra during the school Christmas break. They had not told their parents about the plan, and they did not have a plan of what to tell them. They needed to concentrate on what they were going to do once they got to Petra.

Jill and her father picked Sam up the day before Thanksgiving Day and headed for Buffalo, New York, to celebrate as a joined new family. Jill's dad was excited to go to meet Sam's parents, and Sam's parents were happy to have Jill's dad join them for the holiday as well as anything else that might be announced.

It was a long drive for Sam and Jill, because they had a lot on their minds to talk about but could not talk about it with Jills dad present. They had to spend hours talking about class, teachers, and school events, rather than ghost stories.

Jill's dad wanted to talk about their plans, such as what they were going to do when they finished their graduate work. He asked about where they were going to live and work. Of course, he was always trying to push Jill into living in Boston and getting a job in teaching at one of the many universities in the area. To change that subject Jill simply added to it. "You know, Dad, there is something real important Sam and I want to say to you and his parents. But we want to wait until Thanksgiving dinner for that. Right, Sam?"

Sam looked in the rearview mirror from his rear seat where he could see Jill's face as she drove her dad's Mercedes through New York. "Yeah, right! Mr. John Hart, we would like to sit down with you and my parents to have a long talk. We …"

"Okay, Sam, my dearest, that's enough. Save the rest for dinner. We don't want to spoil your mom and dad's time," Jill said as she gave Sam a silent snarled through the mirror.

"Oh, okay, I'll save the rest for dinnertime."

"Oh, is this not great weather we are having, and the trees are just so beautiful through this part of New York," Jill said, as her dad added.

"I can't wait to meet your parents, Sam."

"Yes sir, Mr. Hart," Sam said as he settled back in his seat and out of sight.

"John! Sam, please just call me John."

"Yes Sir, after we get settled in at my house, John I hope to chat with you about some medical issues."

"Oh, Sam is there something bothering you? Gosh, have you been injured or ill?"

"Oh, no sir, I just have been doing a lot of research for Professor Steinhart on his as well as my projects and they all boil down to human…well life." Sam said as he darted slowly back and forth to avoid the aiming rearview mirror.

They arrived at Sam's house late in the day. His mom and sisters and their husbands had been preparing for the next day big family dinner. The house was swarming with kids and family. There was not much time for Sam and Jill to talk things over or a chance to be alone. Jill and her dad shared a small bedroom while Sam took the couch. Finally, about midnight, as everybody scattered to find their beds, Jill told her dad she was going to kiss Sam goodnight.

Her dad stepped over Sam's sleeping bag, which Jill was borrowing for the night. He climbed into bed and said, "Okay sweetheart, just keep it a kiss and get back here, it's getting very late."

"Yes, Father," she replied as she went out the door and to the living room where she found Sam already asleep. She gently shook him awake.

"What … what is it?" he asked.

"Sam, what are we going to tell them at dinner? Now they all think we're getting married."

"I don't know, Jill. This was your idea. You are the one who announced you had something special to tell them. So, you'd better come up with something."

Jill sat quietly for a moment up next to the couch and asked, "So—is there nothing you would like to say to them?"

"No, I do not have anything to tell them yet. It will be another few weeks before I can say anything."

Jill slowly got up. She looked down at Sam's face and said, "Okay—I will come up with something. See you in the morning."

"Okay, Jill, talk to you in the morning."

Jill said goodnight to her dad and climbed into Sam's sleeping bag. She could tell it still had not been washed after their long journey. It still had the smell of spice. The aroma of clove reminded Jill of Sam's dorm room as well as the first few structures they came to in Petra. The big difference was that it was no longer zipped together with her matching North Face bag.

The next morning, the house was filled with the hustle and bustle of fixing Thanksgiving dinner. Sam spent most of his time talking with his older sisters and their families. Jill helped in the kitchen while her dad watched TV.

Sam's house was small yet had a never-ending backyard that stretched into the woods. Jill decided it was time for a walk in the woods, so she invited her dad along with Sam and his mom and dad to join her and Sam on a walk. As they all left the house headed out together, there was not a window on the backside of the house that did not have at least one face glued to it. Just as they all got in the covering of the woods, Jill turned around and said, "Dad, you remember I said that you might see me reading about or watching something on TV about ghosts?"

"Yes, I do. So, what does that have to do with today, my dear?"

"Well, it has everything to do with today. You see, Sam and I are going back to Petra during Christmas break. We have got some unfinished business to do there, and I know you're not going to like it."

Sam's mom put her hand up in the air, saying, "Wait, wait, wait just a minute. Why is it you are speaking for Sam in this? Moreover, what is the reason for a return to Petra? Especially at Christmas, which is a big family time here as Thanksgiving is. In addition, I hate to interrupt your plans, but somewhere money falls into this plan, and that is not something we have got for another adventure. Sam, what is this trip?"

"Well, well, yes, it is a trip that we need to take. We have some unfinished schoolwork we both are working on."

"So, what does that have to do with ghosts?" Jill's dad intervened.

"Nothing," Sam said.

"But Jill just said something about ghosts, and I have got a sneaky feeling there is something here we are not all sharing."

"Dad, it does. I am trying to prove the ghost stories of Petra are false, and it has to do with some archeology going on there."

Sam's dad finally stepped in as he leaned up against a tree. "Well now, maybe there is something I can help you out with there. I have just taken on a hobby of proof theorems and am having a great time with them."

"No, Dad, this is not a time for your math hobby or mine."

Jill's dad walked over to a big tree stump, sat down, and began playing with a big maple leaf he had picked up. Slowly, he said "Kids, Sam, and Jill, let's stop the runaround and get down to the seriousness of this trip or whatever it is that is you're holding out on us. Jill, you know as well as I do that, I will be funding this trip as I did the last. I did not mind.

I do not mind helping you out with school anytime, but let's have it. I am not getting up from this here stump until you spit it out, all out."

Jill looked at Sam, who took a seat on a fallen log with his elbows on his knees and his face in his hands. She took a deep breath then let out a long sigh and then the whole story as quickly as she could. After everyone was shivering a bit, and afraid to speak they decided to all go home. It was a somewhat quiet walk home, with only a few comments. The three parents looked at each other with long somber looks. Jill and Sam walked back side by side with a great relief on their faces.

Sam's mom just shook her head as she walked over to Sam, put her arm around his shoulders, and said, "Well, I guess you are old enough, like your dad says, to make your own decision. I just ask you and Jill to look out for each other and be extremely careful. And I must say that I still cannot believe this story. It is something that just has never existed, and I could not expect anyone to believe it. I cannot wait until your return and as the old saying goes; *this is out of your system."*

"Yes, this is not a believable story, and I have a good mind to go with you." John said. Then he added, "But I am not saying you two are lying. I just have got, to see what you have or come up with after Christmas. It had better be something for real, or you will have to dream up a car of your own because I will be holding yours for ransom until your return. Sam, I will hold you responsible as well, because I am paying your way since it may very well be my daughter's idea. You, however, from the sound it, will make history or have to go hide somewhere in a few months when the cat is out of the bag.

Jill finally scooted over, put her arm around her dad's waist, and told him how thankful she was for him and his understanding. As they approached the back of the house, the faces were no longer glued to the windows, but one could be seen taking a peek every few seconds. When they entered the house, nothing but smiles and surprised faces could be seen. Everyone was ready for the great surprise. However, as the five grim faces entered, the surprised faces deflated like a tight filled balloon released in a large room.

Thanksgiving dinner was a bit quieter than most thought it would be, but all accepted the story that Sam and Jill had announced to their parents that they were breaking up after three years. Everyone went on with the big family gathering and wished Sam and Jill the best of luck getting back together or finding another. It really wasn't anything each and every one of them of them had not been through themselves before. The next day, John, Jill, and Sam headed back to the Boston area. John became more interested in their backpack adventure and their ghost encounters. He had many questions, some not so serious and some particularly serious. Sam got to air out many questions he had on the doctor's advice on a newly developing human. In his conclusion, he told Sam that what he was doing was just about exactly, what takes place in any animal's womb. However, a bit faster. Sam sat back in his seat and said, "Yes and something most would say is impossible, but it's not. It's that, it is such a difficult unconventional process that keeps people away from accepting it. However, it really, is just like helping a new individual human life begins, with two cells (which both already have life) join up. Then it's all up to DNA and RNA in the right environment."

Sam felt he had all that in place and simply cheated science out of a lot of time by finding the, souls Energy of life, which he found in the electrons. Just before they got to MIT, John asked Sam, "So, with the great experiment you're doing, there must be a lot of complications in preparing the formula and the apparatus to conduct it. So how did you come up with that and are you going to be able to keep it all hid?"

"Well, this type of experimenting has been going on at MIT for years, so it was not hard to come up with the equipment and the chemistry to rebuild a human structure. Several laboratories around the world have been growing life, if you want to call it that, such as animal organs. The big difference between what they are doing around the world and what I am doing is that I am dealing with a human life not an experiment. I will keep it hid the best I can—for a while anyway."

"So, who in the world came up with the great formula for this?"

"For what I am doing, well, it's various ideas many open and unrestricted thinkers out there have been working on for years. I am a lucky one to find one key element. The hard thing about it is that events in our worldly nature can go disarray; but it's pretty simple when you just look at it in-depth."

"So, what does it take, a three-inch three ring binder to keep it in?"

"No, it is really kind of simple. I am just taking the elements or matter contained in a human structure, including the energy that is found in these elements, and then instead of waiting for lighting to strike, I just add the soul of life which carries the code and energy, and *voila*, a human being . Jill and I have simply been introduced to a lost soul in life, that is a ghost that has the DNA code of a particular person, which we all have, and I am just helping it rebuild itself."

"Okay, Sam, sounds nice, but you know me through my daughter, Jill. I am a skeptic, not an old stick-in-the-mud traditionalist. I just want to see, hear, and feel the evidence. You two have a great story, but now you must prove it to me."

"Well, just after Christmas, when we return you will get just that—proof.

When Sam was let off, he did not go to his room. Instead, he made a beeline for the botany lab where he could check up on the project. All was well. The instruments showed the rapid rate of cell growth and reproduction, but no signs of heart rate or respiration. Sam sat down beside the incubator and unplugged the headphone cord that he had piping music into the incubator to help Haifa distinguish day and night. He rolled his thumb around the iPod ring until he found the one, he liked—"With or Without You" by U2. Listening to the gathering beat of the drum made Sam want to sit there until she had a beat of her own. The rhythm made Sam's heart pulsate faster and faster. He had to concentrate to prevent his heart rate from exploding. He knew at this point that she must be reborn, or he could not live with the outcome. He wanted to aid the world out of its narrow-mindedness and help the knowledge of his generation into the future. Now he had another purpose in his life. That was to help another regain the life she lost. After listening to the song until he fell asleep, he awoke finding himself in darkness. The school had been basically shut down for the holiday and was empty. The music had stopped because it was night. He reconnected the headset cord, locked the place, and went to his room. He needed to get a good night's rest and work on the Youssef lyrics some before classes started back to confirm with himself that he could get Jill close as possible to finding the tomb.

Jill spent the night with her dad and then drove back to school early the next morning. There she caught up on all her classes and went over the information Sam gave her on the castle site. It would be her chance to show the people at Petra and her graduate-school advisor that she could discover a particularly important feature at a site through old written documents. She would like to be the first person to announce to the world that she also unraveled the mystery of ghosts. However, that would have to wait. She was also overly concerned as to why Raul was so interested in this finding. After some studying, she found that she could not concentrate on much of anything until she called Raul.

Sunday was a good day because that was the first day of the week and all were back to work. Raul had good news for her and was expecting her and her friend in a couple of weeks. He said that he had made arrangements for them to stay at the Cleopetra Hotel, as they had before. When they arrived, they would show him where the tomb of Katrina was. He said that she would be allowed to use Petra workers to do any digging or excavating. As was the international rule, nothing would be allowed to leave the country, with one exception. That would be one artifact given to her Jill Hart for her assistance in a significant find at Petra. She and Raul would share the credit for the discovery. The current archeologist would not be involved and knew nothing about this upcoming adventure.

Jill was excited but cautious. Raul was very congenial to Jill and seemed too welcoming. She called Sam to give him the good news. Sam's mind was in another world. He did not seem to even want to talk to Jill. When she asked about the Youssef lyrics, he simply explained to her that he had a diagram drawn out for her. He roughly explained to her what he thought the words meant.

He said with a bit of frustration, "Jill, here it is from the top. Katrina had been put to rest in a place other than a place like everyone else in Petra. All people were buried inside the facade structures built throughout the valley, but Katrina had not been buried this way. To get to the tomb, one needs to go to the plaza grounds in front of the castle, which was her house. Find the giant pineapple that was painted on the six-sided stones. Then, within the pineapple, find the one stone that has something in common. Mathematically this is three Fibonacci numbers of 5, 8, and 13, which are in the golden ratio, (ϕ) of the golden rectangle. All pineapples show diagonal lines going through a common skin patch on the hexagonal patches. This common stone in the plaza will be where they would need to enter an aqueduct. Then going under the ground of six heading south toward the castle they would end up somewhere under the house floor. They would then go up through the floor, which would put them at the head of the snail. When we go up into the castle, we can start to eliminate the squares in a golden rectangle that is the shape of the castle floor. On the inside floor where we come up will be the largest square, which will be the snail's head, when depicted on a golden rectangle. The smallest square will be the snail's tail. That is where we will have to dig and go down to find the tomb. There, I guess, we will find the remains of Katrina and something gold in color. So, are you ready to go?"

"Wow, that does make more sense than before," Jill said, adding, "All is ready but you, Sam. You do not sound ready. How are things going? Is there any way I can help?"

"No, there is nothing you can do. I will call you in a week."

The only thing Sam needed was time. Hopefully, that need would end about a week before Christmas. Then his need he would have to deal with the outcome. He had been making several visits a day to the botany lab as well as his workstation, alongside the Professor as well as taking classes. Not only was Sam losing sleep, but he was also losing weight. He was showing the wear and tear of his endeavor. Late one afternoon, he entered the locked room and took his normal seat where he could read the incubator's computer data and listen to his favorite music. The first thing he always did was to turn the iPod music down if it were playing so he could also monitor the computers speakers. As he reached for the volume control, he thought he heard something strange outside the room's door. He got up quickly to look out the covered room window for visitors.

Outside noise was not an unusual thing, because there were other students and professors who used the lab throughout the day. From the door, he could see no one. However, he could still hear something. It was the computer. Not only did he have it responding to several vital-sign detectors in the incubator, but it also monitored sound. He moved back to his chair in front of the computer monitor, not touching thing he turned his ear to the computer's speaker, he could hear a faint rhythm. Quickly, he grabbed the iPod and turned it off. It was not U2 drums, but the beat continued. It sounded like time, *tick-tock, tick-tock, tick-tock*. It *was* time. He moved his hands to the keyboard and started pulling up data. Sam's face lit up like a five-year-old at Christmas. All of the vital signs showed—not only could he hear the heartbeat, but it was also working and circulating blood throughout the human structure. Life had begun to stir again. However, it was not time to open the box. According to the oxygen and carbon dioxide level, it would be another week or so before he could safely open the incubator's lid.

Sam called Jill that night and told her that he would open the incubator the very next weekend. He asked Jill if she wanted to join him that day to see the arrival of Haifa.

Jill was glad that all was working well but had mixed emotions. She did not want to be there if something was not right. She felt the dread that most soon-to-be mothers felt—that was having one born without ten fingers and ten toes. She also dreaded seeing the perfect other girl. She said to Sam, "I will be busy that weekend studying for two tests. I will just see you at the airport the day after we get out of school." With hope as well as hesitation Jill asked, "Are you going to need another ticket?"

"Yes, I will need a ticket for my sister. I have her passport and other identification. They are going to work fine as far as age, size, and things like that. We just might have to use a little makeup."

"You think of everything, don't you? Why don't you ever screw up like I do? I always blame you for things that go wrong and the way you say things. But your mind never shuts down, does it. Sam, I do hope all works out, and your 'sister' will be almost perfect. So, call me the day you open it. Let me know how I can help."

"Okay, I'll call next Sunday. In the meantime, pack your bags."

The following week went by slowly for Sam. He could hardly stay away from the botany lab. However, when Sunday arrived, he had to force himself into the lab. The computer showed vital signs, as it would with any human being. He started to open the lid, but stopped in his tracks, thinking, *she will have no cloths—she will be naked.* He stopped what he was doing, grabbed his keys, and ran to his car.

About thirty minutes later he returned with his extra bed sheets and a light blanket, his mom always sent with him. Now he could open the lid and cover Haifa as a newborn. He quietly unlatched the eight locking latches and broke the sterile seal. As the lid was raised, Sam could smell clove. He could smell Haifa. As he placed the sheet over Haifa, he could see her toes moving as he stretched the sheet down over her knees. There was extraordinarily little unused body material left. His calculations were remarkably close. He could feel the warmth of her body as his hand touched her knee. He turned his head to follow his left hand holding the sheet just below her shoulders. As he lowered it, Haifa's face came into view. Her almond-shaped eyes were open and looking at Sam. They were dark brown and partial shielded by long eyelashes. Her hair was the same as the day her body expired. Her black hair was long and fine, with a slight ribbon wave.

Her eyes closed. The sheet over her chest went down and stopped. Sam could see no movement from her uncovered toes to her brown eyes. A buzzer went off at the computer monitor. Sam did not have to go to it to see what had happened. He could see and feel that she just quit breathing.

Without thinking, Sam pushed his left arm under her neck and grabbed her left shoulder with his right hand. With a rapid motion, he pulled her up and at the same time, her weight pulled him downward. He covered her mouth with his and instantly pinched her nose. He blew a deep breath into her chest and felt it rising as it touched him slightly. He felt his lips turn cool as he lifted his head away from hers a bit. She exhaled, and then he gave her another breath. When he lifted his head, he thought to himself that he wished he could become part of her and would breathe for her forever. Again, she exhaled. As he lowered his mouth to hers, he felt air whisk past his lips. He raised his head, and, from the corner of his eye, he could see her eyes open again.

Again, without thinking, he asked, "Are you all right?" Only a slight smile came across her face. He pulled her to his chest and just held her. Her arms moved to wrap around his chest but were weak and could hardly hold on to him.

Sam whispered, "Haifa do you know me? Do you know what happened and where you are at?"

Sam felt a nod of her head and an incredibly quiet, "Yes."

He let loose of her slowly and lowered her into the incubator, asking again, "Are you all right?"

This time, a smile spread across her face, showing her pearl-white teeth perfect as any young girl. She began moving her feet, legs and reaching for Sam, but had little strength.

Sam asked her if she wanted to try to sit up, and she nodded her head in a *yes* fashion. He realized she was weak and the only thing that would give her strength now was food. He had not thought of that before. Sam realized that she wanted out of the incubator as she held her arms out. "Can you eat?" he asked.

Smile after smile lit up her face as she said, "Drink first. I am just like a baby, I think."

Sam reached for his coat and said, "I'll be right back. I will go get you a milkshake. It will not take me long."

As he turned to the door, he heard her say, "No, no, don't leave me."

He stopped dead in his tracks again, finally thinking, *Wow, how can I leave her. I can't ... What am I going to do?*

He turned to her and said, "Wait right here, don't try to get out. I am going to go start my car and let it be warming while I run back here, then I will carry you to my car and take you with me. Sam was now wishing he had been forced to endure the special high school project of taking care of the baby. He did not have to because he was an AP student, but now he realized that being in the Mensa High IQ society did not give one the knowledge of childcare.

She gave him the okay head nod, and he took off to start his car. One or two wandering students could be seen and only at a great distance. He grabbed an extra coat from his back seat and went back to Haifa. She had push and pulled herself up into a semi-reclined position. Sam helped her put his coat on, disconnected the computer, and shut the incubator down. Then he wrapped the semi-wet sheet around her and lifted her up. Her muscles were like that of a newborn, developed but uncoordinated. She wrapped her arms around his neck and held on as he carried her out to his car. She was able to retain a semi-seated position, with the seat reclined back a little. Sam drove to a Wendy's just down the street and ordered a chocolate shake for Haifa and a vanilla shake, double cheeseburger, and fries for himself. Then down to the nearest parking spot at his dorm. He had to wait a few minutes for the coast to clear, and then he made a mad dash carrying Haifa to his room. He laid her on his bed and returned to get their food. Haifa could not eat but about half of a shake but moaned with delight with each bite. By late afternoon, she began to talk more and would not let Sam out of her reach but for a few minutes at a time. He got her to get into a shirt and pants of his and kept her covered up and in his bed. He did not want her to get sick or have any problems he could not solve. He did have Jill's dad as the nearest back up and would not hesitate one second in using him. He had a long talk with him on the susceptibility Haifa would have when she arrived and what to do. He made it truly clear that he was not a true believer in Sam's story or his project. However, he also made it truly clear that he would be their doctor with no charge and would even make house calls. Haifa and Sam sat and talked for hours. As they sat leaned against the headboard Haifa could not stop smiling or keep from turning her face to Sam to listen or to talk.

Sam said to her, "Gosh Haifa you are as bubbly as a fresh opened bottle of champagne even though you are very weak. I cannot imagine what you are going to be like in a few weeks."

Even though she was very bubbly, she was very calm and soft when she talked. She grabbed Sam's for the first time and said. "Sam, we have been talking a lot, but I have not told you thanks, So, I want to say now simply, I thank you so much for what you have done for me and for my family. I cannot not help from looking at you and feeling warm inside. A smile seems to spread through my entire body. I am becoming tired now and feel that I am going to fall asleep. If I do Please don't leave me here alone. Sleeping as a ghost was so easy but now, I am a person again and I am a bit afraid."

"Don't worry I might get out of bed for a minute or two to get a drink or go to the bathroom but that's it. Oh, uh—by the way if you need something or you know, have to uh—you know go to the bathroom, let me know and I will see if I can help."

"Sam, you must be prepared to help me with a lot of things. I know a lot from seeing, hearing in this world but very new at all else. Therefore, you will have to be forgiving and patient with me. Now good night and thank you again."

Soon after Haifa slid down into the covers and feel asleep, Sam's phone rang. It was Jill. She sounded a bit worried or concerned because Sam had not called or text in over a week.

Sam filled her in on his unbelievable day and said that he was so preoccupied that he forgot about calling. He informed her that he would need an additional ticket for his sister. He explained how it all went, that Haifa had been reborn with no major problems, and that she was just as she was as a ghost but in person.

Jill was a bit heartbroken, but deep down in her heart she was also extremely glad at what had taken place. She was ready to meet Haifa in person but was also ready to return to Petra to fill in the voids she felt within herself.

Chapter 7
Back to Petra

The three met at LaGuardia Airport and went through the channels to board the plane to connect them on to Tel Aviv, Israel. Haifa could now get around just about like anyone else but was slow. She literally could not move as fast and accurately as most other people, but then she really did not want to. She was too interested in taking in everything she could.

Jill became a different person after she met Haifa as a human being. They were opposite in many ways, Jill with long slightly curly blonde hair and Haifa long ribbon black hair. Jill was a naturally hard-headed person, somewhat like her dad. She was also kind and always helpful to those in need of help, as well as helping to steer people in need of help away from those who take advantage from those in distress. Haifa was bubbly. She was one to absorb anything going on. Like Sam, she could read a book in a few hours and give you a full synopsis of it in detail. However, they all three could also just throw the book up into the air and ran to the top of the nearest mountain just to see the other side. On the long flight over the ocean to Amsterdam they had two seats together and one alone. Jill and Haifa got the two together. Haifa and Sam were inseparable up until now. Jill and Haifa became bonded like two long lost sisters that had a natural magnetism. Their biggest difference was their soul bond with Sam. In addition, to their shared feelings and ambitions, Jill accepted that with open arms.

When they landed in Tel Aviv, Israel, they had a short wait before boarding the next plane to Amman, Jordan, Jill came to Sam, threw her arms around him, and said, “Sam, you are the neatest person I have ever met, and we are going to be friends for a long time. In addition, Sam, Haifa is so much like you, that I believe there is never going to be a boring moment around you two. When we get back home, you two have got to spend some time with my dad. That way Haifa can get a good medical checkup and my dad can get a really good check up on life as the three of us have, okay?”

"Sure, I think that I could learn a lot from him. I also got a feeling he will have more than that to checkup on."

The next stop was in Amman, followed by a long cumbersome bus trip to Petra. The three of them went straight to the Cleopetra Hotel and checked in. They could only afford one room based on their John Hart grant, so the girls got the two beds and Sam took the floor. They were in their room no longer than the time it took to open up their backpacks and begin unpacking a few things when a knock came at their door. When Sam opened the door, he stood face to face with Raul. The man was alone, but no doubt had been told exactly when they arrived. Sam invited him into their room. For once, they nicely greeted and introduced themselves. They also exchanged their regrets for losing Youssef. Haifa temporary performing as Sam's sister had to really work hard, blocking her mind when it came to talking about her dad. She had been around Raul all his life and as her dad did, she did not care for the man. She or her family had no problems with the man when they were all ghosts. Things changed when her dad took over the physical body of Youssef.

Raul had known the old Bedouin ever since they were young boys. He said to them, "When we got older, we went our separate ways, seeing each other less and less each year. But, while I stayed out in the public, my friend Youssef, became exceedingly difficult to find, he even stayed away from most of his family. And yes, I even accused Youssef of hiding something from me and many others. This is because; Youssef had come into more money than anyone could do, herding goats, and sheep. He did not brag about it or flaunt it, but he suddenly had money to live without stress unlike me. Once he told me that he had found some gold coins while out wandering around in the hills with his herd."

Jill interrupted, "So that is the main thing you are after, gold and or treasure."

"Of course, who would not be? I believed Youssef did know the location of the king's tomb, and that is where he was getting the gold. Of course, Raul said that he was not interested in the gold but the entire discovery for the sake of his country and Petra. This would put him up for a great promotion, and he was willing to help Jill also receive a shared recognition.

After listening, Jill asked, "But Mr. Raul, are you not afraid of the ghosts that, according to stories, may be in the tomb?"

"That is exactly what they are: stories. That is, just ghost stories. The authorities also like these stories because it keeps grave robbers out of the park and the surrounding hills seeking the stories of riches."

"However, you know as well as I know, that you do suspect the presents of ghosts. You have talked with one or two remarkably close friends about this possibility. The problem is you, as no others in this world have the proof. But you would love to have it wouldn't you?"

Sam spoke from the relaxed position of Haifa's bed. "Okay, so when do we start? We will need some help to lift a very heavy stone or two. And when can we go out to the castle area?"

Sam tried to end the duel over who gets what and why. He was faced with a personal challenge in solving the puzzle to just get to the tomb. Haifa sat quietly by concealing the greatest emotion and was the one with the biggest interest.

Raul continued to stare at Jill and said, "First, I must know, what is it you want from the tomb, and where is it we will begin?

Jill said, "I want to share the discovery of the tomb, in writing and I want one artifact I can take back to the school and have put in the museum with my name on it and a letter from you and the senior authorities of archeology here in Jordan giving me credit for finding the tomb with your assistance."

Sam added, now we are getting somewhere as he rolled over on his side, propping his head up on his hand, "We will find the tomb in the castles. That is where we will begin"

Jill sharply turned to him with a glare to shut up.

"Can you tell me now, while looking at me, that Youssef never told you about gold?"

Jill walked over to Raul, looked right in his face, and said, "Youssef never said the word *gold* or spoke about anything of value other than the value of life. We are not interested in gold, so it is all yours. We could never get out of this country with ten dollars' worth of gold. Just let us show you the tomb and let us see it. That's all—and one small item."

Raul looked over Jill's shoulders at Sam still stretched out and Haifa sitting on the foot of her bed. "And you, my friend?"

Sam looked at Raul and gave him the thumbs-up sign. When he looked at Haifa, she said, "Whatever she says. I just came along to see a ghost."

Sam and Jill both looked at Haifa with a big surprise look then both looked at Raul. Jill said, "Yea, she is a paranormal ghost chaser back home and wanted to see what she could find here. You'll just have to ignore her."

"Yea, so if you see her talking to herself, really ignore her because she thinks she can talk with ghosts."

Raul looked at Haifa then slowly rolled his eyes to Sam then to Jill and said, "Okay, I will tell her if I think I see one. I must return to the office now to prepare for our expedition tomorrow. I will see you at the park gate in the morning at seven o'clock, before people come."

Raul left, and the three got their room ready and chatted wildly about their plan for the next day. At dinnertime, Aali was all question and ears, but got nothing but that they were here to work on a contract to come back next summer. Jill kept her eyes on the door and finally got what she was looking for. That was Imad, who she had contacted to bring Saud and Latif to the hotel. She made the plan not knowing if Haifa would be coming. In addition, she could not come here again without seeing Latif. Now that Haifa was with them that feeling was more than double. The aroma of cinnamon and nutmeg filled the room. Haifa and Imad stared at each other as they were introduced. It was unknown to Imad who she really was, but it was not, unknown to Haifa who had come into the room. She wanted to go directly to the two bags Imad had set on the floor beside Jill but could not reveal herself with the danger of getting things into a mess. Imad sat and talked with them while they all ate and offered his help. After all, had eaten Sam noticed that Haifa had not. He informed his sister that she must eat, in order to give her energy for the hard day's work ahead of them in the morning. After dinner, Imad went home and left them the two bags.

When they got to their room, Jill opened the bags as a signal for the ghosts to appear. Haifa and her brother were shocked and confused as they looked at each other. They knew very well who each other were but their appearance to each other was different. However, to them, appearance was like most of the buildings they had lived around for years, that is a facade. For them, the inner appearance was what was real. It was like music, you could not see it, but one could feel it as their body absorbed it into the mind and let it flow throughout the body. Sam was beginning to discover this and someone else was beginning to crave it. For the first time in hundreds of years, Haifa cried. Her brother came to her, and they wrapped their arms around each other. Haifa fell to her knees, while her twin brother floated down with her. Jill and Sam joined each other in silence as they allowed the two a mixed reunion. Saud also hovered near by with mixed emotion.

After the warm reunion they all talked and shared their thoughts about what they were about to do. Latif and his sister had a few unanswered questions for each other. Most were from Latif. He was not aware that his sister was planning to leave their home again. He was busy helping Jill while his sister made plans to try and find a solution to their infinite problem with Sam's help. After all feeling were healed Jill looked at Sam and said, "Sam, there is no way these two are going to separate again. I hope you realized that could happen."

"You are most likely right. I came fully prepared. Do you think I could out vote two of you on that issue?" Sam said as he rolled out his sleeping bag and went to the bathroom to put on his pj's.

Jill watched him as he returned and said, "And what do you mean by that?"

"I give up, I am going to bed now, I think you have been talking about that a whole bunch, so good night you two, Oh, and you two also Latif and Saud.

At breakfast, they picked up their sack lunches and informed Aali that they would be spending a few nights to put him as ease, though they had different plans if everything went well. When they arrived at the main gate, Raul and two scrubby looking workers were waiting for them. They all piled into a park truck and drove to the castle. There, they all had to wait in suspense as Sam strolled around the grounds of six, the castle plaza. Sure enough, he found the old, faded outlines of a giant pineapple that had been painted on the ground. He drew it out on his spiral and then drew diagonals around three rows of stones, singling out one that was in all rows.

Sam turned to his followers, pointed to the diagram on his notebook, and said, "We need to remove that stone."

He then walked off examining the castles plaza until he found the stone that matched the one in his drawing. He turned to all following him, and then pointed to the one that needed to be removed. Raul got up off the bench he was sitting on while he watched Sam. He walked over to Sam with his two helpers behind him and abruptly asked, "Why remove this stone out here when the castle is over there fifty meters?"

Sam looked at him and said, "The note from Youssef, which you have read many times yourself, said 'Enter the grounds of six.' Well, you are standing on is the grounds of six, and we are going to enter the ground here. Actually, you do not have to. I alone will be fine."

"I am going with you," Haifa said.

Sam turned and grinned at her, and they both held out their hands, reaching and smiling at each other.

"Okay, moving right along—Raul! Can we get your helpers to lift the stone out?" Jill interrupted.

Raul turned to his helpers giving them orders to get their wheelbarrow load of tools over to the stone and pry it out. The stone was removed, and a tunnel could be seen directly below. Sam and Haifa were lowered down into it. They walked south toward the castle, where they found a split to the east and west about fifty meters from where they had entered the ground. As they walked through the old secret tunnel, they banged on the ceiling so that the others above them could follow along. The banging sound led them directly to the nearest corner of the castle and to what was once the inside. At the intersection below, Haifa and Sam found a niche with a small stairway going up. At the top of the short stairs, they banged on the stone above their heads and hollered that this was the stone the party above needed to mark.

Jill hollered to them, telling them they had marked it. Sam and Haifa then went back out the way they came. Sam went to the marked stone, looked at Raul, and said, "Now we are in the castle."

"Yes, an almost destroyed castle. So where is the snail so we can see its head and tail?" Raul asked.

"We are standing on its head. Its tail will be somewhere over there," Sam said as he pointed to the southwest corner of the ruins. Now I need to count the floor tiles that cover the floor. I just need the width and the length numbers." Raul and his helper stood and watched as Sam counted the squares, followed closely by Haifa, Jill, with her strong scented backpack.

Sam stopped counting and announced, "Yep, just as I thought, the castle has fifty-five squares in length and twenty-one in width. This forms a golden ratio in the shape of a rectangle. From this end of the castle, we will section off thirty-four by thirty-four squares, leaving us another golden rectangle on the south. Then we do the same, sectioning off a twenty-one-by-twenty-one square. That will leave us a twenty-one by thirteen golden rectangle. On we go to the very square stone that will be the snail's tail. Within this fascinating golden rectangle, believe it or not, the curve of many seashells and snails fit." As Sam paced off his squares and recorded his movement on his notebook, he said, "If you don't believe me, someday just draw yourself out a good old golden rectangle, and then lay a snail shell on it and you will see what I mean."

"Sam, I think they believe you and only want to see the key stone removed and to see the tomb, as I do. We can talk about how you figured this out later, but let's move the stone, or hopefully the door," Jill stated.

Raul stepped up to the key stone Sam pointed to and spoke to his two followers. Then he said, "Okay, okay. So, is this the stone to the tomb you are pointing at? If so, please move so that we may lift it."

Sam and his group stepped aside as the stone was lifted. There was a stairway leading downward. Raul was the first to enter, followed by his helpers. As Sam and the others went down the stairs, they heard Raul yell, "There is nothing down here, just a big room with a few stone benches and statues and junk. This is not a tomb."

"It is the portal to a tomb," Jill said. "You know a waiting room. Most all tombs in this part of the world had one or more antechambers. Now, we must find a hidden door leading to the tomb or another antechamber. We need to spread out and scan the walls for a doorway, which most likely is an outline smaller than the wall tile joints."

Everybody fanned out in the large room filled with columns holding the castle floor up, looking for an entrance door.

Haifa whispered to Jill as she stood at the one and only marble wall and rubbed her finger up and down the wall. "Jill, this is the only marble wall here and marble is what the Romans used to tightly seal us in the spice niches. The door or sealed wall to the tomb must be in this wall. Scanning the wall Jill soon detected a color pattern showing the outline of a door or maybe the last sealed stone put in place.

"Here is the door, or just last stone put in place, I think. When you bang on it, it sounds thin compared to the wall around it." Jill announced as she winked at Haifa and Sam standing nearby.

Again, Raul butted in with his helpers and their tools. Without questioning or discussion, they began rapidly hammering and chiseling away at the faded-out line. The door soon gave way and easily pushed back on a rail system into the new room.

Raul quickly ordered his helpers to push the door completely into the unknown area as he stepped back. When the sealed door was pushed back about a foot, the seal was broken, and a strong musty smell gushed out. They pushed the door to its end, making an opening to the left and right. Raul ordered the helpers to look inside. When one announced that, in his light, he could see treasure, he was quickly pulled back by Raul, who stepped in and announced, "Ah, I have found it—the great tomb of the last Nabataean king who led the world out of building pyramids over themselves and replaced it with the begging of the castle era."

Inside the small room, two sarcophaguses could be seen resting on two large marble platforms. As Jill, Sam, and Haifa looked over the sarcophaguses, Raul explored the treasures along with his delighted helpers.

As she scouted the small room Jill announced, "Hey, there is a tunnel leading out of this room, and it looks as if the door was opened years ago. But I only see two sarcophaguses, one with a man's supine body carved on its lid and the other with a lady. I do not see Katrina."

Haifa and Sam stood next to the feet of the two side-by-side buried people. As Haifa placed her hands on the lids and rubbed them, tears flowed down her cheeks.

Sam put his arm around her shoulders and rubbed her arm. He whispered to Haifa, "Your grandfather and grandmother aren't they."

Haifa nodded her head and whispered back, "Yes, my grandfather and grandmother."

Jill joined Haifa and Sam, placing the spice bags on the sarcophaguses. She hugged Haifa and said, "I do not see your mother, but there are some …"

Raul interrupted, telling them that he was going to the entrance room to pull the floor stone over the hole so that no one would discover what they had found until they had looked it over good.

Jill then finished, "Some bones are scattered around in the far corner over there," and pointed behind them. At the same time Jill finished talking, and as Raul left the room, a smile spread across Haifa's face. She softly said, "She is here."

Jill stepped back and said, "Where, where is she?"

Haifa closed her eyes and smiled, saying, "Right here. She is standing right before me—my mother. A whisper could be heard but not understood near Haifa. Haifa then said, "She says that her mother and father were killed by the Romans, but they allowed Nabataean slaves to bury them here. Before the tomb was sealed, she hid herself in here to rid herself of the Romans forever. She died in this room, leaving her soul trapped, but she does not understand exactly what took place." Haifa then pointed to a far corner where there were a few bones scattered around and added, "She says that those bones are what is left of her body."

Jill asked Haifa, "Does she speak only in her Nabataean language? Has she never been outside? And ask her what happened to the other door and—why are her bones scattered around and some broken?"

"I cannot see her, but I can feel her. She is holding my hand right now. She has not been exposed to the outside world, so she only knows her first language. She said that a tiger clawed its way in looking for food. Some of the animals were trapped in the tunnels that were the passages from the Lions Tomb structure to the Roman theater where the Christian leader presented Nabataeans to be killed and eaten by tigers and lions in a drama for their entertainment. When the Romans enlarged the amphitheater, they destroyed many Nabataean houses and tombs as well as tunnels. This destruction left all trapped beneath our home."

As Sam listened, he pointed his headlamp at the area of bones, pointed at something, and said, "Jill, look, there is a golden object laying among the bones. It looks like a gold bracelet."

Jill turned and then walked over and picked up the object. As she walked back to Haifa, she held it out to her and asked, "Is this your mother's? Is this a bracelet?"

Haifa reached out for it and said, "Yes, it is my mother's. Inside is a lock of my hair as well as a lock of Latif's from when we were babies. It is a common style of bracelet worn among the Nabataeans."

Sam quietly said, "*just before that which you once did hold.* The bracelet was before her hand. Well, there is proof of Katrina. I guess we should go now and leave the rest to Raul. I think that would be your one item to take home." Jill and Haifa both looked at Sam with a flustered look and Haifa said, "Sam my dear, give me your hand! I would like to introduce you to my…"

Suddenly, they were all startled by the loud sound of gunfire. It sent a shocking reminder through Jill's body. Several shots could be heard coming from the anteroom soon followed by a light coming into the room. The four had been too occupied with Katrina to realize Raul had left them. When the light entered the burial chamber, it quickly was turned to their faces blinding them. As their eyes adjusted to the incoming light, they could see Raul standing at the head of the two sarcophaguses. He held his flashlight on them with his left hand and a handgun on them with his right hand and said, "In order to keep a secret, one must not have any witnesses. So, I have just eliminated two, and now I must add three more."

"You bastard!" Jill yelled out at him across the tomb, and then she started to charge around it to get at Raul. Raul raised the pistol, aiming first at Jill.

At the same time, he squeezed the trigger he yelled and dropped the pistol as well as his light. He fell backward onto the hard floor, grabbing his wrists in pain.

Sam charged around the other side, grabbed Raul's gun and light, and threw them in the entrance room—without thinking he could have used the gun on Raul.

Seeing where his gun and light went, Raul slowly began recovering himself and dashed for the dark room, to retrieve what was shocked out of his hands, by an invisible force.

Haifa's voice then came from the broken door saying, "Quick, over here!" Sam turned and ran to Haifa. Jill suddenly felt a familiar tingle and was led to the sound of Haifa. With their three headlamps pointed down the dark tunnel they all ran, following Haifa as she said, "My mother, I mean our mother, said to go to the end of this aqueduct where it ends at a large underground cistern. She also said to not stop, because there are other ghosts down here."

"Why did she say, 'Do not stop'? Are the ghosts not good? And who hit Raul?" Sam asked as he ran behind the two girls.

"We both shocked him," Latif said. "Saud and I had been out of our bags and in the dark area of the room where you could not see us. My mother asks if you did not see the claw marks on the sarcophaguses. She says the other ghosts are a tiger and a lion that were also trapped below the castle. She says that they do not like people who are not Nabataean. She also says they smell like Saud and me. When they are asleep, they have the aroma of cinnamon and nutmeg. That is because they died while they lay in a large basket of that spice. Our mother cannot be seen I think because there was nothing for her soul Energy to gather in.

Sam had estimated in his mind that they had run at least a football-field length when he slowed down and shined his light down a side corridor. He saw something kind of glittering and said, "I think I just saw one forming down that last hallway, so don't anybody stop or even think about it." He then heard something behind him. Without slowing down, he turned his head looking behind him. He could see another light not far behind. It was Raul hot on their trail. As Sam turned back around to tell the others, a shot was fired.

Raul yelled out, "Run all you want, but there will soon be an end in this damn place, and an end to you."

Haifa came to a sudden stop a few feet from the edge of the large underground water cistern her mother had told her about. She then told everyone what her mother said to do. "My mother says that in front of us is a water-pressure door. To each side of it there is a walkway completely around the wall of the deep cistern. This door in front of us is now in the up position. The doors are operated by weight caused from water pressure on them. There are two water-inlet doors, one to our right and one to our left, each halfway around the side of the cistern. This water tank was built to allow fresh rainwater to flow in through the side doors. When the tank became full, the pressure on this door in front of us and the one across from us was forced open allowing water to flow to the castle and out the other way. At the same time, the inlet doors closed to keep out trash and unwanted critters and remained closed until the water level dropped to a certain level. She says that this door should go down when the weight of two people stand on it. That will allow us to run around to the other side and to escape through the fourth door. However, it will then go down to the water level, which is about ten feet below us."

Time was running out. Raul had come to a walk, knowing he had them trapped.

Sam pushed Haifa toward the walkway to the right on told her to take her mother, go over to the door, and head for to the other side. With an encouraging push, he told Jill to follow them taking Latif and Saud with her. After they got past the door, which did not go down because there was not enough weight on it, they ran around the inner walkway until they came to the inlet door, which was still down and had a large of a gap to jump across. Sam then made the motion to follow them, and then suddenly stopped on the door, which moved slightly. He then turned and faced Raul. Putting his hands up in the air, and said, "Let them go, I know what it is your looking for and can tell you where it is at."

"I already have what I was looking for. There is nothing more you can tell me." Raul said as he pointed the gun at Sam's head.

"You do not know where the larger amount of gold and treasures are in the hills of Petra that Youssef told me about."

Raul walked up to Sam as he stood at the doors edge closest to the water's, and said, "What a gentleman you are. Now give me your light." As Sam unfastened his light, he turned it toward the dark cistern. The water surface was about ten feet below and so motionless Sam could see his reflection. He raised the light up from the water's still surface, shined it to the opposite door, which the beam hardly found. Then he scanned it past the side door, where he could see Jill and Haifa approaching a gap soon to be filled by the rising door. He turned to Raul as the door they were standing on began to shutter downward. Just as Sam handed Raul his light, he dropped it at his feet. As the door went down with, he and Raul's weight the side doors were going up, allowing the others to cross the wide opening, yet leaving Sam stranded. Raul quickly turned to run back down the tunnel they all just came from. He had to jump and grab the floor as the door went down. As he held on the door stopped moving because of the lost weight. Sam looked across the water and could see the lights of Jill and Haifa directly across from him. They had made it to the far side. To avoid meeting Raul at the top Sam jumped into the dark water below and swam for the lights. When his body weight left the door, it began going back up. Raul was saved as the floor came back up under him. When it came to a stop, he ran around the walkway to catch the others and wait for Sam or just let him drown. He soon discovered that, with the first door up, the side doors went back down. From the inlet gap, he took aim on Jill and Haifa who moved onto the fourth door, which began descending again along with the first door, protecting them from Raul's sight. Not able to shoot at them he turned his sights onto the water surface searching for Sam. When he spotted Sam, he took aim and fired a couple shots splashing water on Sam, but not hitting him. His third shot sparked off the ceiling as his light beam began twitching every direction. A loud roaring and snarling filled the air, giving Sam a bit of encouragement.

Haifa began crying and yelling for Sam. Jill also yelled. They did not know he had jumped into the water until they had seen Raul's beam on him. They now had their light on him as he approached the door they were standing on at the water's edge. Haifa kept her beam on Sam, as he got closer. Jill turned hers to the now screaming noise going on between the first and side doors. As the screaming came to an end, she could see a glittering large tiger smelling and pawing at something on the ground. As the tiger began to quiet down, they heard the smooth splashing of water. Jill turned her light to join Haifa's; Sam reached the doors edge and began climbing out, with the help of Jill and Haifa. Before he had his complete stance, Jill and Haifa were both hugging him and thanking him for helping them.

With a gasping voice, Sam said, "Not now, the side door is up. The tiger can cross, Jill; you're the strongest, so climb up on my shoulders and pull yourself up the side and into the tunnel leading out of here, then turn around and reach for Haifa who will be standing on my shoulders as you get out. We have to get the weight off to get the door up trapping the tiger on the other side of the inlet door."

Jill franticly climbed up on Sam's shoulders, grabbed the edge of the floor, and pulled herself up. She then turned her back into the tunnel and reached down for Haifa's hands. As Sam pushed Haifa up and Jill pulled on her small hands the door began rising. When Haifa was completely out the door began, rising faster and Sam could grab the floors edge. However, they were late in getting the door up allowing the tiger along with his friend a ghostly lion to great them as the door became flush with the floor. As Jill and Haifa grabbed each other, they turned their light beams on the ghostly cats staring at them a few feet away. Surprisingly, the cats did not charge them as they did another unfortunate person.

While still on his knees, Sam softly said, "Okay let's quietly move into the tunnel and try not to encourage them into a fight."

As Sam stood, he grabbed Haifa's hand, which was reaching for him. He also met Haifa's face as she bent down and put her palm on his cheek saying, "My mother say's we are fine. We do not have to run. She say's these ghostly cats will not bother us."

"Are you sure?" Sam asked.

"Yes of course, I am sure; after all, my mother has lived with these ghosts for two thousand years. I think they are friends now, and I think she speaks their language. She says that this tunnel before us will lead to an exit. She knows it well but unfortunately as ghosts we don't have the power to dig and move dirt."

"You have the power to open locks and iPods, but she could not move the dirt." Jill added into the talk as all calmed down.

"Yes, that is because our electron flow caused a vibration to allow the lock tumblers to move to their least resistance position. Or if it is electronic, we can speak its language." Haifa answered.

"That is why we feel a vibration when a ghost touched us." Sam added.

"Are you saying Haifa has lost her vibrations, now that she has become a reborn human?" Jill asked Sam as a grin spread across her face.

Sam quickly recovered. "No way, but more importantly I am learning to understand my own with two great teachers."

Haifa's mother said goodbye to the two cats after they re—operated the doors allowing them back into their underground home. Then they followed the tunnel, which began going slightly up hill. It had several water inlets that had been sealed off by the Romans as well as the earthquakes over the years past. At the end of the main tunnel, they found themselves against a dirt roof.

"My mother says that just above here is the building called the Lions Tomb and that we can dig ourselves into the floor here. She says that like all the structures carved into the mountains it is all rock, except here where the dirt fills one of the many niches inside the tomb."

After about an hour of digging, they found light. It was the dim light from the inside of the Lions Tomb. When they got the hole big enough, they dragged themselves out and refilled the hole. Then they crawled through the tiny space between the top of the outside door and the many years of fallen rock and dirt, that had almost covered this structure up.

When they all reached the outside and finished giving each other hugs and thanking each other for their freedom, Sam turned to the faint carvings of the lions or tigers on the structure's facade and said, "Wow, I see why they call it the Lions Tomb. I am glad they were there. I owe them special thanks, for taking care of that traitor, Raul. That gave me time to swim to you guys. I just wasn't thinking far enough ahead that they would come after us. Therefore, I am really glad we had their best friend with us. Now let's all head to the truck and see if we can get out of here without being noticed." As they walked toward the castle, they chatted like a bunch just getting off the Mission SPACE ride at Walt Disney World.

"I wonder how many people Raul told about this adventure we all went on?" Jill said.

"I don't know, and I hope none, but we will soon see when we get to the castle. I hope your old friend the park archeologist is not there," Sam said as he and Haifa led the pack, arm in arm.

Haifa asked, "Was Imad coming into the park to pick us up? He did know where we were going, didn't he?" she asked, looking back at Jill as she drifted behind everybody.

"Yes, I told Imad what we were doing and what would be involved, from the information I could get from Sam. If he is not there at the castle waiting for us, we should find him at the hotel. I hope he is there waiting; we need to get out of here today and head for Amman."

Sam and Haifa both grinned at each other, as they almost appeared to be skipping. Sam said. "Yes, we are done here and ready to go back."

"We, did you say we? Who all are we?"

"Well … me, Haifa, Latif, their mom Katrina, Saud and you are of course."

"Sam, have you all, already talked about this and know what will be involved? I mean, where are they going to stay, and is this what they want?"

"It is what I want. I am sure it is what Haifa wants, and like you said, there is no way they are going to be separated for a while now."

From behind Jill, Saud said, "I will not go with you at this time. Maybe some other day you can come back for me. While you people have been talking, my sister and I have also been talking. She wants to go, and so does Latif."

Everybody stopped and turned around to look at the cloudlike humans Saud and Latif.

Jill said, "What, your sister?"

"Yes, I did not wish to tell you before, but Katrina is my younger sister. When the Romans took her away, they did not want any more, royal trouble than the King and Queen to deal with or to start trouble. I was to work as a slave or be entombed with the other rebellious Nabataeans. As you see, I was entombed in a spice room until I seeped out like the others. We both thank you for bringing us together. You know, she has never been exposed to your language or your kind of people or anything like the rest of us, so she would like to go with you and get away from here. And Jill, I think Latif would like to talk with you later, but he also wants to go with his mother and sister."

Jill ran her palm across her forehead, turned around, and headed off to the castle, which was in sight now and said, "Very well, let's head for home. Man—O—Man, is my Dad ever in for a shock."

"Yeap he sure is. He is going to have to open a new practice for ghosts' physicals and post reborn physicals, for sure." Sam said as they approached the castle.

"What's that for and what all do I have to do for that, Sam." Haifa asked.

Jill now in the led stopped and locked her arm into Haifa's free arm and said, "We will talk about that on the long flight home. Not much else was said other than Sam telling of a plan to get Katrina and Latif on the plane and what proposal they could make with Dr. John Hart when he picked them up at the airport. Sam needed to prove they had spent the Doctor's money appropriately. Jill needed to explain her new discovery as well as introduce a couple new friends.

They found Imad sitting on the park truck tailgate. He had brought his goat truck just in case they would need a different one. They choose the goat truck leaving the park rangers abandoned where he left it. As he drove them to the Cleopetra, Jill explained all that had taken place and some of what they found. Imad could not find where they had entered the floor of the castle because Raul had replaced the opened tile. The park was carrying on as in a normal day. The park archeologist was never told about anything that was going to take place. Raul had to keep it all a secret in order to gain what he really wanted, and the secret was now known to just those found in Imad's goat truck. They gathered up their stuff and checked out within thirty minutes, leaving Aali scratching his head again. Imad drove them all straight to his uncle's house in Amman, where they spent the night, said their farewells, and parted ways.

Chapter 8
Home at Last

After three long connecting yet tedious flight to LaGuardia, Jill's dad, who was in for a big surprise because of such a fast trip. He met the three plus at the arrival gate. Jill ran to her dad and grabbed him around his arms with a bear hug burying her face in his chest. Crying she said, "Dad I am so glad to be home and I think you are going to have a hard time with this story."

"Why, what is wrong, did you have bad trouble?" he asked as he glared at Sam.

"No, Dad, all went very well, but we have some work to do now here at home and we really are going to need your help in more ways than one." Jill released her Dad, looked into his eyes, and added, "Let's get to the car and we will begin to explain what we can." Just after they got in the car, Sam said, "Now, Dr. Hart, you asked for proof of our ghostly experience."

"Yes, Sam, I did, but would you first introduce me to your new friend who appears to have taken my daughter's place?"

"Dad, she did not take my place. Sam and I had been kind of splitting up naturally, and yes, he found … well, I think he found his soul mate. I am also incredibly happy for him and her."

"O—kkay, my dear, I think I am getting at what's going to be hard to understand. So please tell me who might this be?"

"No Dad, the hard part is yet to come."

Sam looked around at everybody and then said, "Dr. Hart, I would like for you to meet Haifa. She is the one I was working on in my little project that you were not told had arrived. She was a ghost from Petra and now is a regenerated human being or born-again soul … whatever, she is just Haifa, oh, and she needs a physical."

John came to a stoplight, put his arm over the seat, and turned around, looking Haifa over.

Haifa put her petite hand out to John and said, "Hello, Dr. Hart, I am glad to meet you, and I thank you for funding this adventure. I hope that I am not too much of a surprise for you and a problem."

John twisted to look at Sam by her side and then at Jill. He said, "This is not much proof. In my counseling career, do you know how many 'werewolves,' 'vampires,' and 'ghosts' I have met or seen and talked too? Sam and Jill, you have got to come up with something better that that."

"Dad, it's a green light, let's go."

"Yes, my dear. Just got a little sidetracked there."

Sam scooted closer to Haifa, leaned over the seat a bit, and said, "I figured that. Okay, how about if I could show you a ghost—that is, to see, feel, smell, hear, and even taste if you want. Would you believe me then?"

"Yes, Sam, you have a deal there, but I will have to be the one who makes the final judgment."

Sam looked over at Jill who just continued to stare out her window and tried to stay out of Sam's deal. He then turned back to John and said, "Okay, one more thing. If I give you the evidence, and you accept the proof using your scientific method, will you give a couple of them room and board? They don't eat much."

Jill quickly turned to Sam and said, "Sam, what on earth are you saying? That is a crazy idea. That just won't work. Sam …"

John laughed a bit and then said, "You have a deal, Sam. I will gladly do anything you want or give you and you collaborate here, my quiet daughter, anything you need if you can show me a ghost. When do I get to see these spices of life?"

"When we get to your house, you can meet them."

Jill turned back to her window and put a pillow up between the seat and door so she could take a nap. She closed her eyes and said, "I think we found you one, Dad. Wake me when we get home." Her Dad gave her a blank stare then returned to the busy traffic he was facing.

When they got to Jill and her dad's house, they unloaded everything, which amounted to just two backpacks and an extra bag of gifts and souvenirs.

As they unpacked in the big laundry room, John walked in, slapped his hands together, rubbed his palms briskly, and said, "Okay, where are my ghosts, kids?"

Jill said, "Dad, you'd better sit down. I am not sure how your heart is going to take this."

"I'll be fine. How many are there? Now you're going to tell me they're invisible, right?"

"Dr. Hart, there are two—one is totally invisible and the other is visible and will pass all your tests. But first, I want you to meet Katrina. So, say hi to her first."

"Okay, Ms. Katrina, hi, how are you?" John said as he grinned really big and then put a straight face on and looked at Jill.

Jill shook her head slowly and said, "Dad, just look straight ahead and take about two large steps."

John turned his face forward, looked down at the floor, took one large step, stopped, and then took another large step. After his first foot hit the ground and just before his second foot got halfway to landing, he closed his eyes and gasped, "Oh my God, oh God, what is that?" His knees buckled under him as if he had just been caught in an electric fence while trying to sneak between the wires. As he gathered himself from falling to the floor, he was able to step back. He shook his head and then his whole body shimmered. As he shook his hands, he said, "What in the heck was that? I could hear someone talking in a strange language. I could see stars or something like that." He looked at Jill and then at Sam who both were sitting calmly watching. "That's it? Was that it? I did not really see anything. I am not sure what it was, but it was something."

Sam said, "Try it again and just listen. Haifa will translate for you because her mother whom you kind of, entered there does not speak English. This time, just hold out your hand, and she will take it."

Haifa reached out and took one of John's hands as he slowly raised it while she reached into the air just in front of John's hand for Katrina's. As she put the two together, John quickly asked for a chair. He sat down as Jill scooted a chair under him. As he looked at the three standing before him, and said, "When I stepped into her, it was like seeing a star off in space, at first, but then the star came zooming in very quickly. As it got closer, it began to sparkle more and suddenly it burst into millions of other stars like if I were flying into a small compact galaxy. I was surrounded with nothing but sparkle and warmth. She also made me tingly, inside, and out. Wow, now I have heard her talk. Not sure, I want to taste her. I smell a spice, so I think I am convinced." John sat back, still breathing a bit hard as if he had just run a 5K.

Jill said, "That's not Katrina you smell, she has no physical features other than those one can feel. The essence of nutmeg you smell is that of Latif. He is her son and Haifa's twin brother. He is appearing right before you from that little souvenir bottle of spice on the floor."

As John looked down at the little bottle, he saw a light brown cloud start to rise up out of the bottle, quickly forming into a young man looking at Jill. Within seconds, Latif floated toward her, turned to John, and introduced himself.

As he held his hand out to John, the doctor asked, as he squeezed his hands together, looked at Jill, "Will it be the same?"

Sam laughed a bit and beat Jill saying, "Not even close Dr. Hart, not even close. They have a way to channel their feelings and their intensity."

John took Latif's hand and said, "Welcome, welcome to you and your mother to your new home away from your old home." After meeting Latif, he turned to Sam and said, "We have a lot to talk about." He then walked over to Jill, wrapped his arms around her, and said to her as tears flowed from his eyes, "I did not believe you. You of all have now proved to me that the soul exists, and it can be standing right in front of you and you cannot see it. You cannot see it for your closed mind."

Jill hugged her father strongly and said, "Dad, there's this place a little bit south from here I think you might want to go see. It is called Machu Picchu and has a great hotel right near one of the most mysterious places on earth. You think you might want to go see it?"

"Are there ghosts there?"

"Well, maybe. We just need to check out a story I heard. Also, we now have a ghost with us. I think a ghost could find another ghost if there is one, so what do you think."

Sam said, "Well let's do some work here first. Maybe we can convert another ghost or two, and then maybe we can pack our bag!"

Made in the USA
Monee, IL
04 November 2026

67379961R00108